D1112168

BIOMECHANICS AT MICRO- AND NANOSCALE LEVELS

VOLUME I

BIOMECHANICS AT MICRO- AND NANOSCALE LEVELS

VOLUME I

editor

Hiroshi Wada

Tohoku University, Sendai, Japan

NEW JERSEY • LONDON • SINGAPORE • BEIJING • SHANGHAI • HONG KONG • TAIPEI • CHENNAI

Published by

World Scientific Publishing Co. Pte. Ltd.

5 Toh Tuck Link, Singapore 596224

USA office: 27 Warren Street, Suite 401-402, Hackensack, NJ 07601

UK office: 57 Shelton Street, Covent Garden, London WC2H 9HE

British Library Cataloguing-in-Publication Data
A catalogue record for this book is available from the British Library.

BIOMECHANICS AT MICRO- AND NANOSCALE LEVELS
Volume I

ISBN 981-256-098-X

Printed in Singapore by Mainland Press

PREFACE

A project on "Biomechanics at Micro- and Nanoscale Levels," the title of this book, was approved by the Ministry of Education, Culture, Sports, Science and Technology of Japan in 2003, and this four-year-project is now being carried out by fourteen prominent Japanese researchers. The project consists of four fields of research, which are equivalent to four chapters of this book, namely, Cell Mechanics, Cell Response to Mechanical Stimulation, Tissue Engineering, and Computational Biomechanics.

Our project can be summarized as follows. The essential diversity of phenomena in living organisms is controlled not by genes but rather by the interaction between the micro- or nanoscale structures in cells and the genetic code, the dynamic interaction between them being especially important. Therefore, if the relationship between the dynamic environment of cells and tissues and their function can be elucidated, it is highly possible to find a method by which the structure and function of such cells and tissues can be regulated. The first goal of this research is to understand dynamic phenomena at cellular and biopolymer-organelle levels on the basis of mechanics. An attempt will then be made to apply this understanding to the development of procedures for designing and producing artificial materials and technology for producing or regenerating the structure and function of living organisms.

We are planning to publish a series of books related to this project, this book being the first in the series. The trend and level of research in this area in Japan can be understood by reading this book.

Hiroshi Wada, PhD,
Project Leader,
Tohoku University,
Sendai,
March, 2004.

CONTENTS

I. CELL MECHANICS

IMAGING AND MECHANICAL PROPERTIES OF GUINEA PIG OUTER HAIR CELLS STUDIED BY ATOMIC FORCE MICROSCOPY

H. WADA, M. SUGAWARA, K. KIMURA, Y. ISHIDA, T. GOMI AND M. MURAKOSHI

Department of Bioengineering and Robotics, Tohoku University, 6-6-01 Aoba-yama,
Sendai 980-8579, Japan
E-mail: wada@cc.mech.tohoku.ac.jp

Y. KATORI, S. KAKEHATA, K. IKEDA AND T. KOBAYASHI

Department of Otorhinolaryngology, Head and Neck Surgery, Tohoku University, Graduate
School of Medicine, 1-1 Seiryo-machi, Sendai 980-8675, Japan

High sensitivity of human hearing is believed to be achieved by cochlear amplification. The basis of this amplification is thought to be the motility of mammalian outer hair cells (OHCs), i.e., OHCs elongate and contract in response to acoustical stimulation. Thus, the generated force accompanying the motility amplifies the vibration of the basilar membrane. This motility is concerned with both the cytoskeleton beneath the OHC plasma membrane and the protein motors distributed over the plasma membrane, because it is presumed that the cytoskeleton converts the area change in the plasma membrane induced by the conformational change of the protein motors into OHC length change in the longitudinal direction. However, these factors have not yet been clarified. In this study, therefore, the ultrastructure of the cytoskeleton of guinea pig OHCs and their mechanical properties were investigated by using an atomic force microscope (AFM). The cortical cytoskeleton, which is formed by discrete oriented domains, was imaged, and circumferential filaments and cross-links were observed within the domain. Examination of the morphological change of the cytoskeleton of the OHC induced by diamide treatment revealed a reduction of the cross-links. Results of the examination indicate that the cortical cytoskeleton is comprised of circumferential actin filaments and spectrin cross-links. Mechanical properties in the apical region of the OHC were a maximum of three times greater than those in the basal and middle regions of the cell. Moreover, Young's modulus in the middle region of a long OHC obtained from the apical turn of the cochlea and that of a short OHC obtained from the basal or the second turn of the cochlea were 2.0 ± 0.81 kPa and 3.7 ± 0.96 kPa, respectively. In addition, Young's modulus was found to decrease with an increase in the cell length.

1   Introduction

Even though the amplitude of tympanic membrane vibrations is only a few nanometers when we speak in a low voice, we can clearly understand what is being said. The human ear is characterized by such high sensitivity, which is believed to be based on the amplification of the basilar membrane (BM) vibrations. This amplification is made possible by the motility of the outer hair cells (OHCs) in the cochlea, this motility being realized due to the structure of the lateral wall of the OHC.

The lateral wall of the OHC consists of three layers: the outermost plasma membrane, the cortical lattice and the innermost subsurface cisternae. In the plasma membrane, there are protein motors which possibly change their conformation according to the membrane potential. As a result of this

3

4

conformational change, change in the area of the plasma membrane occurs [1]. The cortical lattice, which consists of actin and spectrin filaments [2-4], beneath the plasma membrane is expected to convert this area change in the plasma membrane into length change in the axial direction of the cell. However, although the mechanism of the motility is concerned with the structure and mechanical properties of the OHC lateral wall, these factors have not yet been clarified.

In this study, therefore, first, the ultrastructure of the cytoskeleton of the OHC was investigated in the nanoscale range using an atomic force microscope, which is a powerful tool for studying biological materials [5-7]. The mechanical properties along the longitudinal axis of the OHCs taken from each turn of the cochlea were then obtained by using the AFM.

2   Materials and Methods

2.1   Imaging of the cytoskeleton of OHCs

2.1.1   Cell isolation

Guinea pigs weighing between 200 and 300 g were used. They were decapitated and their temporal bones were removed. After opening the bulla, the cochlea was detached and transferred to an experimental bath (the major ions in the medium were NaCl, 140 mM; KCl, 5 mM; CaCl$_2$, 1.5 mM; MgCl$_2$ 6H$_2$O, 1.5 mM; HEPES, 5 mM; glucose, 5 mM; pH 7.2; 300 mOsm). The bony shell covering the cochlea was removed and the organ of Corti was gently dissociated from the basilar membrane. The OHCs were isolated by gently pipetting the organ of Corti after enzymatic incubation with dispase (500 PU/ml). The isolated OHCs were transferred to a sample chamber and glued to MAS-coated slide glass (Matsunami glass).

The care and use of animals in this study were approved by the Institutional Animal Care and Use Committee of Tohoku University, Sendai, Japan.

2.1.2   Sample preparation

First, an attempt was made to obtain AFM images of the OHC lateral wall without fixation. However, as the cell wall was very soft, it was impossible to obtain images. The isolated OHCs were, therefore, fixed with 2.5% glutaraldehyde and simultaneously extracted with 2.5% Triton X-100 in phosphate buffer (pH 7.4) for 30 min at room temperature. After fixation, the OHCs were rinsed three times with 0.1 M phosphate buffer solution. The OHCs were then observed by the AFM in liquid.

In some experiments, diazene dicarboxylic acid bis [N,N-dimethylamide] (diamide), which reduces the actin-spectrin binding mediated by protein 4.1 in erythrocytes [8], was used for modifying the cytoskeleton of the OHC. Diamide was dissolved in the bath solution containing dispase (250 PU/ml). The diamide concentration was 5 mM. The dissociated organ of Corti was incubated in these solutions for 30 min at room temperature. After the incubation, the OHCs were isolated by gently pipetting the organ of Corti. The isolated OHCs were then transferred to a chamber and fixed and extracted in the same way as the cells which were not incubated with diamide.

2.1.3   Atomic force microscopic imaging

A commercial AFM (NVB100, Olympus) was used for the experiments. As the AFM unit is mounted on an inverted optical microscope, positioning of the tip above the cells is easy. V-shaped silicon nitride cantilevers (OMCL-TR400PSA-2, Olympus) with a pyramidal tip and a spring constant of 0.08 N/m were used. The typical radius of the curvature of the tip was less than 20 nm. In this study, as OHCs were too soft to resist lateral friction force during scanning in the contact mode, the oscillation imaging mode of the AFM (Tapping Mode™, Digital Instruments, Santa Barbara, CA, USA) was used. In the tapping mode AFM, the cantilever tip oscillates and touches the sample only at the end of its downward movement, which reduces the contact time and friction force as compared with the contact mode.

In this experiment, the frequency of the cantilever tip oscillation was between 3.8 and 5.2 kHz, which is close to the resonance frequency of the cantilever tip. The scanning regions were 0.5×1.0 μm and 1.0×1.0 μm. The scanning frequency was fixed at 0.4 Hz (scan speed: 0.4 μm/s). In all AFM images, the sample was scanned from left to right. The scanning direction corresponds to the axial direction of the OHC.

All images were analyzed by a software program by Digital Instruments (Santa Barbara, CA, USA). To correct dispersions of individual scanning lines and remove background slopes, images were plane fitted and flattened. After that, to show the fine structure more clearly, the contrast of the original AFM images was enhanced using the software program. The surface profiles were obtained by the section analysis of the original AFM images, and calculation of spacing between adjacent filaments was done by the same procedure as reported previously [9].

2.2   Mechanical properties of OHCs

2.2.1   Cell preparation

Guinea pigs weighing between 200 and 300 g were used. OHCs were isolated from the animal by the same procedure mentioned in Section 2.1. With this isolation procedure, OHCs from the apical turn, those from the third turn and those from the basal or second turns could be classified. However, the row of the cell could not be distinguished. All experiments were performed at room temperature.

6

2.2.2 Indentation test by using the AFM

Figure 1 depicts the principal of an indentation test by using the AFM. When the cantilever is moved by a piezoelectric scanner and the tip of the cantilever comes in contact with a sample, the cantilever starts to deflect. The deflection of the cantilever is detected by an optical detector which is composed of a laser, a mirror and a photodiode array. By this measurement, the relationship between the movement of the cantilever z by the piezoelectric scanner and the cantilever deflection d was obtained. This obtained curve is termed force curve because the force applied to the sample can be calculated by multiplying the cantilever deflection d by the spring constant of the cantilever. In each measurement, the force curve obtained

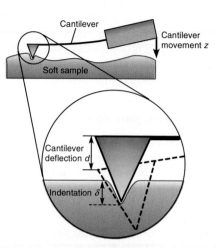

Figure 1. A schema of the cantilever on a soft sample. When the tip of the cantilever is touching the soft sample, indentation δ is defined as the difference between the cantilever movement z and the cantilever deflection d.

when the sample was pushed against by the tip was confirmed to correspond to that obtained when the sample was retracted from the tip, which means that the sample was elastic during measurement. Force curves were obtained at four points at intervals of 20 nm in the circumferential direction, and the mean and standard deviation of the slope of the square regression line or those of Young's modulus were calculated at each point on the OHC.

2.2.3 Analysis of force curves

An example of force curves obtained from both soft and hard samples are shown in Fig. 2A. As the cantilever deflection d is equal to the cantilever movement z in the hard sample, the sample indentation δ is given by

$$\delta = z - d \tag{1}$$

From Fig. 2A, the relationship between the cantilever deflection d and the sample indentation δ is obtained as shown in Fig. 2B. The curve in this figure is fitted with a square regression line which is given by

$$d = a\delta^2 \tag{2}$$

where a is the slope of this curve, which represents the elastic properties of the sample.

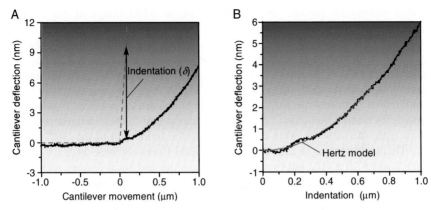

Figure 2.   Relationships between the cantilever deflection, cantilever movement and indentation.   A: Relationship between the cantilever deflection and cantilever movement, i.e., force curve, obtained from both a hard sample (substrate) and a soft sample (OHC).   The red dotted line shows the results obtained from the hard sample.   The black solid line shows the results obtained from the soft sample.   In the case of the hard sample, the cantilever deflection remains steady at zero until the cantilever touches the sample, at which time it rises proportionately with the increase in the cantilever movement.   The vertical arrow shows the indentation, i.e., the difference between the cantilever deflection of the hard sample and that of the soft sample. B: Relationship between the cantilever deflection and indentation. The black line shows the experimental data.   The red line represents a square regression line fitted by Eq. (2).

When subjects are elastic, isotropic, homogeneous and semi-infinite and when the indentation tip is rigid and conical, the Hertz model, which describes the elastic response of subjects indented by the tip, can be applied to the measurement data. According to the Hertz model, the relationship between the cantilever deflection d and the indentation δ is defined as follows:

$$d = \{2E \tan \alpha / \pi k(1 - v^2)\}\delta^2 \qquad (3)$$

where E, α, k and v are Young's modulus of the sample, the half-opening angle of the cantilever, the spring constant of the cantilever and Poisson's ratio of the sample, respectively [10, 11].   In this study, the half-opening angle and the spring constant of the cantilever were 17 degrees and 0.08 N/m, respectively.   Poisson's ratio was assumed to be 0.499 because the samples were biomaterials.   Since the slope a in Eq. (2) corresponds to $2E \tan \alpha / \pi k(1 - v^2)$ in Eq. (3), Young's modulus of sample E can be obtained from this relationship.

3   Results

3.1   Morphology of the cytoskeleton of OHCs

The cytoskeleton of the lateral wall of the fixed OHC, which was extracted with Triton X-100, was imaged with tapping mode AFM.   Figure 3A depicts the

8

measured rectangular region. Figure 3B shows an AFM image of the cortical cytoskeleton obtained in that region. In AFM images, the brighter areas correspond to the higher regions of the sample surface, and the transverse direction in the AFM images corresponds to the axial direction of the OHC. A schematic of the domains and filaments shown in Fig. 3B is displayed in Fig. 3C. In this image, differently oriented domains are recognized. Within each domain, relatively thick circumferential filaments run parallel to each other and are cross-linked regularly or irregularly by thinner filaments. Such lattices, composed of these discrete domains and two types of filaments, were observed along the full length of the OHC lateral wall. The mean spacing ± S.D. of circumferential filaments were 51.5 ± 9.78 nm ($n = 550$) and 47.0 ± 10.2 nm ($n = 352$) in the middle and basal regions and in the apical region of the OHC, respectively. The difference between the mean spacing in the middle and basal regions and that in the apical region was statistically significant at $P < 0.0001$ using Student's t-test. By contrast, the difference

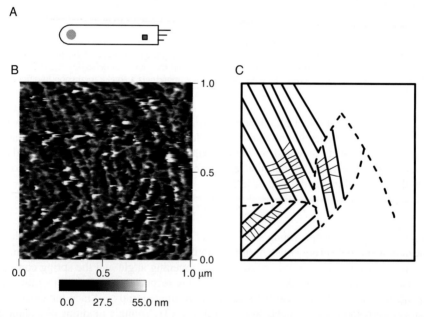

Figure 3. An AFM image of the cytoskeleton in the OHC lateral wall. The OHC was fixed by 2.5% glutaraldehyde and demembraned with 2.5% Triton X-100. A: The position of the scanning area. B: The AFM image. C: A schematic of the domains and filaments in B. In this figure, the schematic only shows clearly recognized areas. Boundary domains, circumferential filaments and cross-links are shown by dotted lines, thick solid lines and thin solid lines, respectively. The cortical lattice is formed by some differently oriented domains. Within each domain, thicker circumferential filaments are cross-linked by thinner filaments. "Reprinted from Hearing Research, 187, Wada et al., Imaging of the cortical cytoskeleton of guinea pig outer hair cells using atomic force microscopy, 51-62, Copyright (2004), with permission from Elsevier."

between the mean spacing of cross-links in the middle and basal regions and that in the apical region was not statistically different. The mean spacing ± S.D. of cross-links was 25.2 ± 7.23 nm (n = 300) along the full length of the OHC lateral wall.

3.2 Effect of diamide on the structure of the OHC cytoskeleton

Some OHCs were incubated with diamide, which is a sulfhydryl-oxidizing agent. The cytoskeleton of diamide-treated OHCs which were fixed with 2.5% glutaraldehyde and extracted with 2.5% Triton X-100 was imaged using the tapping mode AFM. Figure 4 shows AFM images of the cortical lattice modified by 5 mM diamide. Although many circumferential filaments are recognized, cross-links are hardly seen, since diamide treatment reduces cross-links in the AFM images.

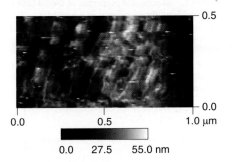

0.5

0.0

0.0   0.5   1.0 μm

0.0   27.5   55.0 nm

Figure 4. An AFM image of the diamide-treated OHC lateral wall. The OHC was fixed by 2.5% glutaraldehyde and demembraned with 2.5% Triton X-100 after incubation with 5 mM diamide. Many circumferential filaments are recognized; however, cross-links are hardly seen. "Reprinted from Hearing Research, 187, Wada et al., Imaging of the cortical cytoskeleton of guinea pig outer hair cells using atomic force microscopy, 51-62, Copyright (2004), with permission from Elsevier."

3.3 Mechanical properties of OHCs

First, mechanical properties of the OHC in the apical turn of the cochlea were measured at various points along the longitudinal axis of the OHC. In this measurement, since the surface profile of the sample in the apical region of the cell could not be regarded as semi-infinite in extent, the Hertz model could not be applied to the measurement data in the analysis. The relationships between the cantilever deflection and the indentation of the OHC were, therefore, measured, and the slopes of the square regression lines fitted to these relationships were obtained. The relationships between the slope and the distance from the basal end of the OHC are shown in Fig. 5. Data were obtained from 10 OHCs. The abscissa represents the distance of each measurement point from the basal end of the OHC, and the ordinate shows the variation of slope, i.e., the value of the slope at each point divided by that in the middle part of the cell. Positions of the basal and apical ends along the cell axis are converted to 0.0 and 1.0, respectively. From this figure, it is found that the slope is almost constant in the basal and middle regions of the cell and that the slope in the apical region of the cell is larger than those in the basal and middle regions. This means that there is no significant difference in the mechanical properties in the basal and middle regions of the OHC. However, the mechanical property in the apical region of the cell is greater than those in the basal and middle regions of the OHC. Moreover, it is clear that the slopes in the apical region of the cell are a maximum of threefold larger than those in the basal and middle regions. To confirm the difference in the slope between these regions, the Mann-Whitney test was used. As a result, in seven of the 10 cells, the slopes in the region between 0.80 and 1.0 from the basal end of the cell were shown to be significantly ($P < 0.05$) larger than those in the region between 0.0 and 0.80 from the basal end of the cell.

Secondly, in the same way as mentioned above, the mechanical property of the OHC in the basal turn or the second turn of the cochlea was measured. Figure 6 depicts the variation of the slope as a function of the distance from the basal end of the OHC obtained from the total of 10 cells. From this figure, it is found that the slopes in the basal and middle regions of the cell in the basal turn or the second turn are almost constant, while those of the slopes in the apical region of the cell are up to three times as large as those in the basal and middle regions. In addition, in five of the 10 cells, the difference between the slopes in the region between 0.80 and 1.0 and those in the region between 0.0 and 0.80 from the basal end of the cell is statistically significant at $P < 0.05$ using the Mann-Whitney test.

When the representative value of Young's modulus of the cell is defined as that in the central part of the cell,

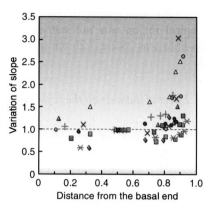

Figure 5. Mechanical properties of the 10 OHCs obtained from the apical turn of the cochlea. The abscissa represents the normalized distance of each measurement point from the basal end of the OHC. The ordinate represents the slope at the point of measurement divided by that in middle part of the cell. The different symbols indicate the data obtained from different cells.

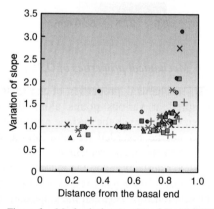

Figure 6. Mechanical properties of the 10 OHCs obtained from the basal or second turn of the cochlea. The abscissa represents the normalized distance of each measurement point from the basal end of the OHC. The ordinate represents the slope at the point of measurement divided by that in the middle part of the cell. The different symbols indicate the data obtained from different cells.

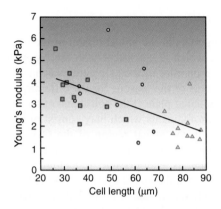

Figure 7. Relationship between Young's modulus and the length of the OHC. Red triangles, green circles and blue squares represent cells in the apical turn, the third turn and the basal or second turns, respectively. The regression line is given by $y = -0.039x + 5.2$ ($r = 0.62$).

from Fig. 5 and Eq. (3), Young's modulus of the cells in the apical turn of the cochlea is 2.0 ± 0.81 kPa ($n = 10$), while that of the cells in the basal or second turns is 3.7 ± 0.96 kPa ($n = 10$) from Fig. 6. As the location of the OHC along the cochlea is related to the length of the cell, the relationship between Young's modulus of the cell and the length of the cell is obtained. The result is shown in Fig. 7, where Young's moduli of the OHCs from the apical turn (red triangles, $n = 10$), the third turn (green circles, $n = 9$) and the basal turn or the second turn (blue squares, $n = 10$) are displayed. From this figure, it is clear that Young's modulus decreases with an increase in the cell length.

4   Discussion

4.1   Structure of the OHC cortical lattice

Diamide is a specific oxidizing reagent for the sulfhydryl group [12]. In erythrocytes, diamide increases intermolecular links between spectrin molecules [13, 14] and reduces the actin-spectrin binding mediated by protein 4.1 [8]. The reduction of cross-links in the AFM image in this study, observed on the lateral wall of the diamide-treated OHC, would be due to the fact that the diamide treatment reduces the actin-spectrin binding mediated by protein 4.1. From immunological evidence, it is known that the cortical lattice is composed of actin and spectrin [4, 15]. This suggests that the circumferential filaments are actin filaments and that the cross-links are spectrin. This idea is supported by the fact that the diamide treatment reduces the axial stiffness of the OHC in a dose-dependent manner [16], which makes the cell highly extendable in the axial direction [17].

A scheme of the OHC cortical lattice is shown in Fig. 8. The cortical lattice is formed by differently oriented domains; such domains consist of circumferential filaments, which are composed of actin, arranged parallel to each other and cross-links, which are composed of spectrin, connected to adjacent actin filaments. In this study, pillars were not preserved, presumably due to the permeabilization of Triton X-100.

4.2   Variance of mechanical properties of OHCs

As shown in Figs. 5 and 6, mechanical properties in the apical region of the OHC are greater than those in the other regions, which is consistent with the report of Wada et al. [18] in which local mechanical properties of the cell were evaluated by applying a hypotonic solution to the cell.   One of the reasons for the difference in mechanical properties might be due to the cuticular plate at the apical end of the cell. As the cuticular plate mainly consists of actin filaments which may consist of hard material [2, 19, 20], mechanical properties in the apical region of the OHC are affected by the cuticular plate and are greater than those in the other parts.   The other possibility is that cell organelles such as endoplasmic reticulum, Golgi apparatus, lamellar bodies and so on are located in the apical cytoplasm beneath the cuticular plate [21].   In this study, as the cell was indented up to 1 μm, which is greater than the thickness of the lateral wall of the OHC, if there were some organelles in the cytoplasm under the cantilever, they could have affected the measurement and the deflection of the cantilever might have been larger.   As a result, mechanical properties in the apical region of the OHC would be large.

As shown in Fig. 7, the length of the OHC differs with its location, i.e., the long cells are mainly located in the apical turn and the short cells are in the basal turn.   Figure 7 also shows that Young's modulus of the OHC decreases with an increase in its length.   Hallworth [22] reported that the cell compliance, which is the inverse of the stiffness of the cell, increased with an increase in the cell length. In order to compare the mechanical properties of the cell obtained in this study with those in his report, Young's modulus obtained in this study is converted into stiffness using a simple one-dimensional model of the OHC.   When the OHC is assumed to be cylindrical and homogeneous, the relationship between Young's modulus E and the stiffness k_{cell} of the OHC is shown by the following equation:

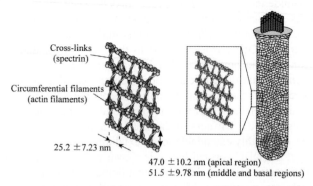

Cross-links
(spectrin)

Circumferential filaments
(actin filaments)

25.2 ±7.23 nm

47.0 ±10.2 nm (apical region)
51.5 ±9.78 nm (middle and basal regions)

Figure 8.   A schema of the cortical lattice.   The cortical lattice consists of differently oriented domains, such domains consisting of circumferential filaments which are composed of actin arranged parallel to each other and cross-links which are composed of spectrin connected to adjacent actin filaments.

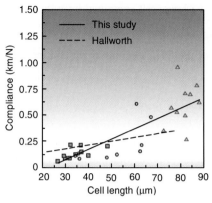

Figure 9. Relationship between the compliance and the length of the OHC. Young's modulus shown in Fig. 7 is converted into the compliance, which is the inverse of cell stiffness. Red triangles, green circles and blue squares represent the cells in the apical turn, the third turn and the basal or second turns, respectively. Solid and dashed lines show the regression line which is given by $y = 0.010\,x - 0.22$ ($r = 0.83$) in this study and that obtained by Hallworth (1995), respectively.

$$k_{cell} = \pi R^2 E / L \tag{4}$$

where R and L represent the radius and length of the OHC, respectively. By taking the radius of the OHC to be 5 μm and substituting the length of each OHC into Eq. (4), Young's modulus of each OHC is converted into the stiffness. As shown in Fig. 9, although the slope is different, both data explain that the compliance increases linearly with an increase in the cell length. As it is considered that mechanical properties of the OHC are strongly related to the force production of the cell [16, 22] and that cells which have small compliance can produce large force [23], our results, therefore, suggest the possibility that the force produced by the OHC in the cochlea might be different along the length of the cochlea.

5 Conclusions

The ultrastructure of the cytoskeleton of guinea pig OHCs and their mechanical properties were investigated by using an AFM. The following conclusions can be drawn:

1. The cortical lattice consists of some differently oriented domains, which are composed of thicker circumferential filaments and are thinner cross-linked filaments.

2. Circumferential filaments and cross-linked filaments are actin filaments and spectrins, respectively.

3. The mean spacing ± S.D. of circumferential filaments were 51.5 ± 9.78 nm ($n = 550$) and 47.0 ± 10.2 nm ($n = 352$) in the middle and basal regions and in the apical region of the OHC, respectively. That of cross-links was 25.2 ± 7.23 nm ($n = 300$) along the full length of the OHC lateral wall.

4. The apical region of the OHC was a maximum of three times harder than the basal and middle regions of the OHC.

14

5. Young's modulus in the middle region of a long OHC obtained from the apical turn of the cochlea and that of a short OHC obtained from the basal turn or the second turn were 2.0 ± 0.81 kPa and 3.7 ± 0.96 kPa, respectively.
6. Young's modulus in the middle region of the OHC decreased with an increase in the cell length.

Acknowledgements

This work was supported by Grant-in-Aid for Scientific Research on Priority Areas 15086202 from the Ministry of Education, Culture, Sports, Science and Technology of Japan, by the 21st Century COE Program Special Research Grant of the "Future Medical Engineering Based on Bio-nanotechnology," by a grant from the Human Frontier Science Program and by a Health and Labour Science Research Grant from the Ministry of Health, Labour and Welfare of Japan.

References

1. Adachi, M., Iwasa, K.H., 1999. Electrically driven motor in the outer hair cell: Effect of a mechanical constraint. Proc. Natl. Acad. Sci. USA 96, 7244-7249.
2. Holley, M.C., Ashmore, J.F., 1990. Spectrin, actin and the structure of the cortical lattice in mammalian cochlear outer hair cells. J. Cell Sci. 96, 283-291.
3. Holley, M.C., Kalinec, F., Kachar, B., 1992. Structure of the cortical cytoskeleton in mammalian outer hair cells. J. Cell Sci. 102, 569-580.
4. Nishida, Y., Fujimoto, T., Takagi, A., Honjo, I., Ogawa, K., 1993. Fodrin is a constituent of the cortical lattice in outer hair cells of the guinea pig cochlea: Immunocytochemical evidence. Hear. Res. 65, 274-280.
5. Wada, H., Kimura, K., Gomi, T., Sugawara, M., Katori, Y., Kakehata, S., Ikeda, K., Kobayashi, T., 2004. Imaging of the cortical cytoskeleton of guinea pig outer hair cells using atomic force microscopy. Hear. Res. 187, 51-62.
6. Sugawara, M., Ishida, Y., Wada, H., 2002. Local mechanical properties of guinea pig outer hair cells measured by atomic force microscopy. Hear. Res. 174, 222-229.
7. Sugawara, M., Ishida, Y., Wada, H., 2004. Mechanical properties of sensory and supporting cells in the organ of Corti of the guinea pig cochlea – study by atomic force microscopy. Hear. Res. 192, 57-64.
8. Becker, P.S., Cohen, C.M., Lux, S.E., 1986. The effect of mild diamide oxidation on the structure and function of human erythrocyte spectrin. J. Biol. Chem. 261, 4620-4628.
9. Wada, H., Usukura, H., Sugawara, H., Katori, Y., Kakehata, S., Ikeda, K., Kobayashi, T., 2003. Relationship between the local stiffness of the outer hair cell along cell axis and its ultrastructure observed by atomic force microscopy. Hear. Res. 177, 61-70.

10. Sneddon, I.N., 1965. The relation between load and penetration in the axisymmetric Boussinesq problem for a punch of arbitrary profile. Int. J. Eng. Sci. 3, 47-57.
11. Wu, H.W., Kuhn, T., Moy, V.T., 1998. Mechanical properties of L929 cells measured by atomic force microscopy: Effects of anticytoskeletal drugs and membrane crosslinking. Scanning 20, 389-397.
12. Kosower, N.S., Kosower, E.M., 1995. Diamide: an oxidant probe for thiols. Methods Enzymol. 251, 123-133.
13. Haest, C.W., Kamp, D., Plasa, G., Deuticke, B., 1977. Intra- and intermolecular cross-linking of membrane proteins in intact erythrocytes and ghosts by SH-oxidizing agents. Biochim. Biophys. Acta 469, 226-230.
14. Maeda, N., Kon, K., Imaizumi, K., Sekiya, M., Shiga, T., 1983. Alteration of rheological properties of human erythrocytes by crosslinking of membrane proteins. Biochim. Biophys. Acta 735, 104-112.
15. Flock, A., Flock, B., Ulfendahl, M., 1986. Mechanisms of movement in outer hair cells and a possible structural basis. Arch. Otorhinolaryngol. 243, 83-90.
16. Adachi, M., Iwasa, K.H., 1997. Effect of diamide on force generation and axial stiffness of the cochlear outer hair cell. Biophys. J. 73, 2809-2818.
17. Frolenkov, G.I., Atzori, M., Kalinec, F., Mammano, F., Kachar, B., 1998. The membrane-based mechanism of cell motility in cochlear outer hair cells. Mol. Biol. Cell 9, 1961-1968.
18. Wada, H., Usukura, H., Takeuchi, S., Sugawara, M., Kakehata, Ikeda, K., 2001. Distribution of protein motors along the wall of the outer hair cell. Hear. Res. 162, 10-18.
19. Flock, A., Bretscher, A., Weber, K., 1982. Immunohistochemical localization of several cytoskeletal proteins in inner ear sensory and supporting cells. Hear. Res. 6, 75-89.
20. Slepecky, N.B., Ulfendahl, M., 1992. Actin-binding and microtubule-associated proteins in the organ of Corti. Hear. Res. 57, 201-215.
21. Saito, K., 1983. Fine structure of the sensory epithelium of guinea-pig organ of Corti: Subsurface cisternae and lamellar bodies in the outer hair cells. Cell Tissue Res. 229, 467-481.
22. Hallworth, R., 1995. Passive compliance and active force generation in the guinea pig outer hair cell. J. Neurophysiol. 74, 2319-2328.
23. Sugawara, M., Wada, H., 2001. Analysis of force production of the auditory sensory cell. In: Cheng, L., Li, K.M., So, R.M.C. (Eds.), Proceedings of the 8th International Congress on Sound and Vibration held at the Hong Kong Polytechnic University, Hong Kong, 26 July 2001.

DEVELOPMENT OF A NOVEL MICRO TENSILE TESTER FOR SINGLE ISOLATED CELLS AND ITS APPLICATION TO VISCOELASTIC ANALYSIS OF AORTIC SMOOTH MUSCLE CELLS

T. MATSUMOTO*, J. SATO, M. YAMAMOTO AND M. SATO

Biomechanics Laboratory, Tohoku University, 6-6-01 Aoba-yama, Sendai 980-8579, Japan
*Current affiliation: Nagoya Insitute of Technology, Showa-ku, Nagoya 466-8555, Japan
E-mail: takeo@nitech.ac.jp

A novel tensile tester has been developed to measure tensile properties of single isolated cells. Both ends of the cell were aspirated with glass micropipettes coated with a cellular adhesive. One pipette was moved with an electrical manipulator to stretch the cell horizontally. The force applied to the cell was measured by the deflection of a cantilever part of the other pipette. The cell was observed with water immersion objectives under an upright microscope to obtain its clear image. Cultured bovine aortic smooth muscle cells were stretched at various strain rate in a physiological salt solution at 37°C. Elastic modulus of the cells had a significant positive correlation with the strain rate, and was about 3 kPa at the strain rate <4%/s. Viscoelastic analysis with a standard viscoelastic solid showed that relaxation time for constant strain of the cells was 164s, which is much longer than those reported for endothelial cells and fibroblasts (40s). Smooth muscle cells may be most viscous among the cells in the vascular wall.

1   Introduction

Biological tissues change their dimensions and mechanical properties adaptively to various mechanical stimuli [1]. Because cells in the tissues play important roles in the adaptation process, it is important to know how much stresses and strains are applied to the cells in the tissue. For this purpose, we need to know the mechanical properties of cells over a wide range of strain From these viewpoints, we have develped a micro tensile tester for isolated cells [2-4].

In this paper, we first introduce you to our tensile tester and show the tensile properties of the cultured bovine aortic smooth muslce cells measured with the tester at various strain rates. Based on these data, we studied the viscoelastic properties of the cells to find that the smooth muscle cells were most viscous among cells related to blood vessels, such as red and white blood cells, endothelial cells, and fibroblasts.

2   Micro Tensile Tester

Figure 1 shows the schematic diagram of the tensile tester. A specimen cell was held by glass micropipettes and stretched with a computer-controlled micromanipulator under a microscope. Deformation process of the cell was observed with the CCD camera and videotaped for an image analysis which was performed after the experiment. Although most of cell manipulation systems have been constructed under an inverted microscope to make cell manipulation easier, we used an upright microscope with water immersion objectives to get clear image of the cells (Fig. 3).

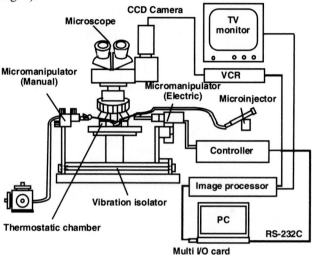

Figure 1. Micro tensile tester for cells.

The details of the test section is shown in Figure 2. The specimen cell was held with two micropipettes, the deflection and operation pipettes, by aspirating the cell surface gently. The tips of the micropipettes were coated with a cellular adhesive (Cell-Tak, Becton Dickinson) to improve adhesiveness between the cell and the pipettes. The operation pipette is connected to an electric micromanipulator and pull the cell to the right. The deflection pipette has a cantilever part which deflects in response to the load applied to the cell and was used to detect the amount of load applied to the cell. Heater, heating block, and a temperature sensor were used to maintain ambient temperature at 37°C.

3   Tensile Test Protocol and Data Analysis

We used bovine aortic smooth muscle cells (BASMs) obtained with an explant method [2] and cultured until 6–7th passage. After tripsinization, the cells were resuspended in Ca^{2+}- and Mg^{2+}-free phosphate buffered saline (PBS) at 37°C and one of the cells was aspirated at both ends with two micropipettes whose tips had

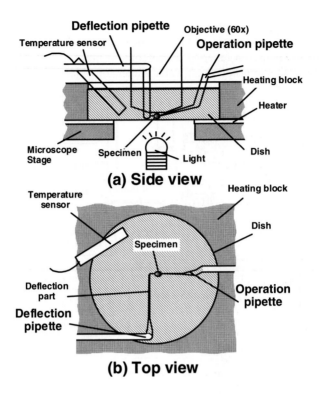

Figure 2. Details of the test section.

been coated with the Cell-Tak. Fifteen to 20 min were allowed to increase adhesion strength between the cell and the pipettes. Then the cell was stretched stepwise with an increment of 1 to 3 μm while the deformation of the cell was continuously videotaped. Time invervals between the steps were 1–10 s. Following the experiment, the deflection pipette was calibrated to obtain its spring constant. Then the cell deformation was analyzed on videotaped images.

We could obtain very clear images of the cells (Fig. 3). We measured the position of the pipette tips to obtain gauge length L and the displacement of the deflection pipette x. The tension applied to the cell F was calculated by multiplying the displacement x with the spring constant of the deflection pipette k which was measured after each tensile test. The elongation ΔL was calculated as the increment of L. The nominal stress σ was calculated by dividing tension F with the original cross-sectional area A_o, which was obtained from the diameter before the stretch D_o assuming circular cross section. The nominal strain ε was obtained by normalizing the elongation with the original gauge length, L_o.

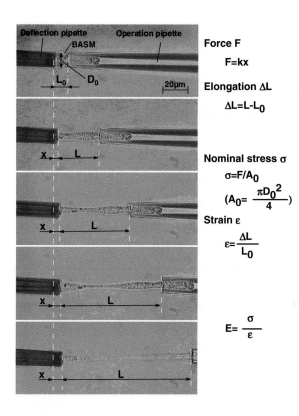

Force F

F=kx

Elongation ΔL

$$\Delta L = L - L_0$$

Nominal stress σ

$$\sigma = F/A_0$$

$$(A_0 = \frac{\pi D_0^2}{4})$$

Strain ε

$$\varepsilon = \frac{\Delta L}{L_0}$$

$$E = \frac{\sigma}{\varepsilon}$$

Figure 3. An example of a cultured BASM during stretch. The image was very clear and even small particles inside the cell can be identified.

4   Results

A typical example of the time course changes of elongation and force of a cell in response to stepwise stretch is shown in Figure 4. The cell showed creep following the rapid elongation due to each stepwise stretch, causing gradual decrease in the deflection of the deflection pipette, and so in the tension applied to the cell. The more the cell was elongated, the more remarkable the creep and stress relaxation became. In the subsequent analyses, we concentrated on force-elongation relationships rather than time course changes. Data at each step was obtained just before next stepwise stretch.

Both of the force-elongation curves (Fig. 5) and the stress-strain curves (Fig. 6) of the cultured BASMs were very sensitive to their strain rates. Most of the cells could be stretched to two times of their original length. However, breaking of the cells did not occur in the central portion of the cells. Some of them broke at the pipette tip and some of them slipped off from the pipette when the strain became large. All the curves shown here were obtained within the time points when the slipping began.

20

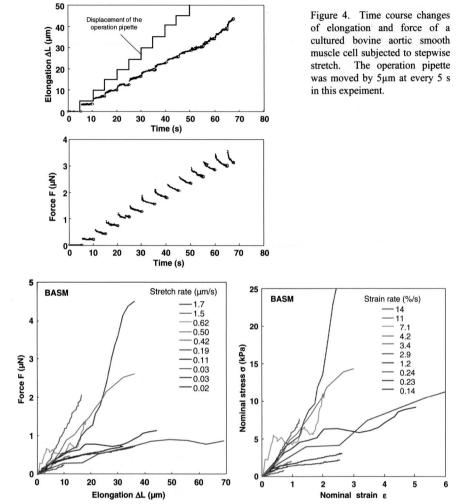

Figure 4. Time course changes of elongation and force of a cultured bovine aortic smooth muscle cell subjected to stepwise stretch. The operation pipette was moved by 5μm at every 5 s in this expeiment.

Figure 5. Force-elongation curves of the bovine aortic smooth muscle cells (BASMs) obtained at various stretch rates.

Figure 6. Nominal stress-nominal strain curves of the bovine aortic smooth muscle cells (BASMs) obtained at various strain rates.

We changed the stretch rate by the factor of 100. It looks like the higher the strain rate is, the stiffer the cell becomes. To check this hypothesis, we calculated elastic modulus from each curve and plotted it against the strain rate. Elastic modulus was obtained by fitting a straight line from the origin to the entire segment of each curve, and was called overall elastic modulus, E_{all}. There was significant correlation between the strain rate and the elastic modulus (Fig. 7). If you increase strain rate from 2 (%/s) to 10 by 5 times, the modulus increase from 2 (kPa) to 6 by

3 times. This indicates that SMCs are viscoelastic material and you need to pay much attention to the strain rate when discussing the mechanical properties of SMCs. Measured parameters are summarized in Table 1.

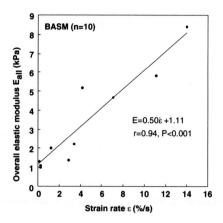

Figure 7. Relationship between the overall elastic modulus and the strain rates.

Table 1. Tensile properties of bovine aortic smooth muscle cells.

Cell no.	Initial diameter D_0 (μm)	Initial length between pipette tips L_0 (μm)	Strain rate ε (%/s)	Spring constant for pipette k (N/m)	Overall elastic modulus E_{all} (kPa)	Initial elastic modulus E_{init} (kPa)
BASM1	17.5	12.2	3.4	0.038	2.2	2.3
BASM2	17.5	12.3	0.14	0.016	1.3	3.7
BASM3	20.5	14.3	0.23	0.013	1.0	3.3
BASM4	12.6	8.8	7.1	0.684	4.7	(16.1)
BASM5	19.5	13.6	0.24	0.002	1.1	1.2
BASM6	12.6	8.9	1.2	0.026	2.0	2.9
BASM7	9.5	6.7	2.9	0.182	1.4	3.3
BASM8	15.2	12.1	4.1	0.100	5.2	4.0
BASM9	18.6	13.5	11	0.380	5.8	3.0
BASM10	13.0	11.9	14	0.157	8.4	5.0
mean	15.6	11.4	4.4	0.160	3.3	3.2
SEM	1.1	0.8	1.5	0.069	0.8	0.4

( ), data not used for initial elastic modulus.

5 Viscoelastic Analysis

We used the standard linear solid or Kelvin model for the analysis (Fig. 8). The governing equation of the model is expressed as:

$$\sigma + \tau_\varepsilon \dot{\sigma} = E_R(\varepsilon + \tau_\sigma \dot{\varepsilon}), \tag{1}$$

where $\tau_\varepsilon = \eta / E_1$ is the time constant at constant strain, *i.e.*, time constant of stress relaxation, $\tau_\sigma = (\eta / E_0)(1 + E_0/E_1)$ is that at constant stress, *i.e.*, time constant of creep, and $E_R = E_0$ is an asymptotic elastic modulus after all viscoelastic deformation diminishes.

By assuming a ramp input of strain $\varepsilon = \alpha t$, you can obtain stress-strain relationship as:

$$\sigma\left(\frac{\varepsilon}{\alpha}\right) = E_R\left\{\varepsilon + \alpha(\tau_\sigma - \tau_\varepsilon)\left(1 - \exp\left(-\frac{\varepsilon}{\tau_\varepsilon\alpha}\right)\right)\right\}. \qquad (2)$$

We determined the parameters τ_ε, τ_σ, and E_R by minimizing errors between the theoretical and experimental curves for each cell.

Table 2 summarizes the viscoelastic parameters of each cell. They are shown in the order of strain rate. It is confirmed that SMCs showed marked viscoelasticity: the time constant of stress relaxation τ_σ was more than 150 s and that of creep τ_ε was more than 1000 s. The identified parameters did not have significant correlation with strain rate except τ_ε and E_1. The stress-strain curves at various strain rates estimated from the model with the mean parameters are summarized in Figure 9. As again confirmed in this figure, strain rate has significant effect on the stress-strain relationships of the smooth muscle cells.

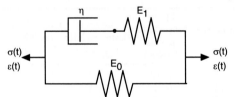

Figure 8. Standard linear solid or Kelvin model used in the viscoelastic analysis. E_0 and E_1, Young's moduli; η, coefficient of viscosity; $\sigma(t)$, stress; $\varepsilon(t)$, strain.

Table 2. Summary of viscoelastic parameters.

Strain rate (%/s)	τ_ε (s)	τ_σ (s)	$E_R = E_0$ (kPa)	E_1 (kPa)	η (kPa s)
14	117	1915	0.48	7.36	861
11	112	1463	0.47	5.62	629
7.1	84	874	0.50	4.71	395
4.2	106	1456	0.47	5.94	629
3.4	388	1106	0.79	1.47	570
2.9	110	788	0.47	2.88	317
1.2	100	1015	0.52	4.80	480
0.24	336	892	0.50	0.83	278
0.23	107	877	0.50	3.60	385
0.14	184	1106	0.79	3.97	731
Mean	164	1149*	0.55	4.11*	527
SEM	34	113	0.04	0.64	60

*$P < 0.05$ vs strain rate

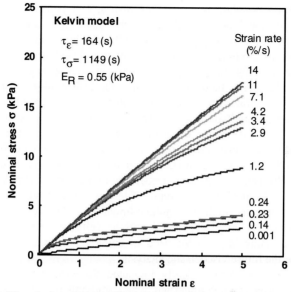

Figure 9. Viscoelastic response predicted with the standard linear solid (Kelvin model).

6   Discussion

We have developed a micro tensile tester for isolated cells and measured the tensile properties of cultured bovine aortic smooth muscle cells.   From mechanical properties obtained at various stretch rates, we estimated the viscoelastic properties of the cells and found that the cultured bovine aortic smooth muscle cells were highly viscous with the time constant of stress relaxation τ_σ being >150 s and that of creep τ_ε being > 1000 s.

Viscoelastic properties of the blood cells and vascular cells are summarized in Table 3 along with those obtained in the present study.   We calculated parameters

Table 3.   Viscoalestic properties of various cells relating to blood vessel.

Cell type	Measurement method	τ_ε (s)	τ_σ (s)	E_0 (kPa)	E_1 (kPa)	η (kPa•s)
Erythrocytes [5]	Pipette aspiration	<0.1				
Neurtophils [6]	Pipette aspiration	0.18	0.65	0.028	0.074	0.013
Endothelial cells [7]	Pipette aspiration	38	116	0.092	0.19	7.2
Fibroblasts [8]	Micro glass plate	40		0.96	0.51	13
SMCs [present]	Tensile test	164	1149	0.55	4.11	527

24

for the standard linear solid from the data shown in each paper. Interestingly, cells flowing inside of the blood vessel have very small time constants while time constant of those in the wall are more than 100 times larger than that of the blood cells. Among them, the time constant of smooth muscle cell obtained in the present study is the largest. This means SMCs are highly viscous among the components in the vasclular wall. Due to their high viscosity, the SMCs might bear much higher stress than other vascular components during pulsation of the blood vessel wall.

There are several limitations to this study. First of all, we used cultured smooth muscle cells for the ease of the experiment. It is well known that cultured cells change their phenotype from contractile to synthetic. We have confirmed that the static elastic modulus decreased to less than one third in response to the phenotypic change [2, 9]. We need to measure viscoelastic properties of freshly isolated cells that maintain their contractile phenotype. The use of the standard linear solid for the viscoelastic analysis might be another limitation. As shown in Figure 6 the mechanical properties of the smooth muscle cells are not linear and their deformation was quite large. We need to employ nonlinear model with finite deformation for the future viscoelastic analysis.

In conclusion, this study clearly indicated that SMCs are highly viscous. Viscoelastic analysis is very important when studying mechanical properties of smooth muscle cells.

Acknowledgements

We thank Mr. Yoshiki Ogawara for his superb technical assistance. This work was supported in part by Grant-in-Aid for Scientific Research (B) 16360052 and Grant-in-Aid for Scientific Research on Priority Areas 15086209 both from the Ministry of Education, Culture, Sports, Science and Technology of Japan.

References

1. Fung, Y.C., 1990. Biomechanics: Motion, Flow, Stress, and Growth. Springer, New York.
2. Matsumoto, T., Sato, J., Yamamoto, M., Sato, M., 2000. Smooth muscle cells freshly isolated from rat thoracic aortas are much stiffer than cultured bovine cells: Possible Effect of Phenotype. JSME Int. J., Ser. C 43, 867-874.
3. Nagayama, K., Nagano, Y., Sato, M., Matsumoto, T. Effect of actin filament distribution on tensile properties of smooth muscle cells obtained from rat thoracic aortas. J. Biomech. (submitted).
4. Nagayama, K., Matsumoto, T. Mechanical anisotropy of rat aortic smooth muscle cells decreases with their contraction: possible effect of actin filament orientation. JSME Int. J., Ser. C (submitted).

5. Hochmuth, R.M., Worthy, P.R., Evans, E.A., 1979. Red cell extensional recovery and the determination of membrane viscosity. Biophys. J. 26, 101-114.
6. Schmid-Schönbein, G.W., Sung, K.L., Tozeren, H., Skalak, R., Chien, S., 1981. Passive mechanical properties of human leukocytes. Biophys. J. 36, 243-256
7. Sato, M., Theret, D.P., Wheeler, L.T., Ohshima, N., Nerem, R.M., 1990. Application of the micropipette technique to the measurement of cultured porcine aortic endothelial cell viscoelastic properties. J. Biomech. Eng. 112, 263-268.
8. Thoumine, O., Ott, A., 1997. Time scale dependent viscoelastic and contractile regimes in fibroblasts probed by microplate manipulation. J. Cell Sci. 110, 2109-2116.
9. Miyazaki, H., Hasegawa, Y., Hayashi, K., 2002. Tensile properties of contractile and synthetic vascular smooth muscle cells. JSME Int. J., Ser. C 45, 870-879.

SHEAR DEPENDENT ALBUMIN UPTAKE IN CULTURED ENDOTHELIAL CELLS

K. TANISHITA, M. SHIMOMURA, A. UEDA AND M. IKEDA

Center for Life Science and Technoogy, Keio University, Yokohama 223-8522, Japan
E-mail: tanishita@sd.keio.ac.jp

S. KUDO

Department of Mechanical Engineering, Shibaura Institute of Technology, Tokyo 108-8548, Japan E-mail: kudous@sic.shibaura-it.ac.jp

To clarify the process of plasma protein uptake, we focused on a negatively charged glycocalyx on the cell surface, since the glycocalyx electrostatically supposed negatively charged protein uptake such as albumin, and the glycocalyx thickness was varied with the variation of shear stress on the surface. After subjected bovine aorta endothelial cell to various shear stress (0.5, 1.0, 2.0, 3.0 Pa) for 48 hours, we determined the glycocalyx thickness with electron microscopy and lead cationic particle, toluidine blue, to bind to anionic charged glycocalyx and measured absorbance of the binding amount with spectrophotometer. We measured the albumin uptake from acquired fluorescent images of fluorescent labeled albumin with confocal laser scanning microscopy at neutralized glycocalyx charge and non-treatment. The albumin uptake on non-treatment, increased at comparatively low shear stress (0.5, 1.0, 2.0 Pa), and decreased at comparatively high shear stress (3.0 Pa). The albumin uptake on neutralized charged cell increased about two fold of non-treatment at 3.0 Pa (P < 0.001). From this study, we found that the glycocalyx thickness and charge were constant at low shear stress, but changed thicker and higher than control at comparatively high shear stress. This result indicates that glycocalyx has the influence on albumin uptake at comparatively high shear stress.

1 Introduction

The atherosclerotic legion appears in the region of low shear stress of relatively large arteries such as the carotid bifurcation and the coronary artery[1]. Since the atherosclerosis was initiated by the uptake of low density lipoprotein (LDL) [2], the uptake of the LDL is highly associated with hemodynamic stress. Some studies demonstrated that the transport of macromolecules such as albumin across the cell membrane is highly affected by the imposed shear stress [3, 4]. Kudo et al. [5] measured the albumin uptake into endothelial cells being affected by shear stress in vitro. Their study showed the increased uptake for lower shear stress and decreased one for high shear stress. However, the mechanism for such biphasic feature of uptake remains unclear, but the endothelial cell interface a key role for regulating the uptake of macromolecules.

The endothelial cell surface coordinate with various extracellular domains of membrane-bound molecules, constructing the glycocalyx. Luft [6] visualized the endothelial glycocalyx layer by ruthenium red staining for an electron microscopic study. They found out the glycocalyx thickness was in the scale of 20 nm. Subsequent electron microscopic observations of the molecules revealed that the glycocalyx thickness is less than 100 nm. However, in vivo study have found thicker glycocalyx layer, in the range from 0.5 μm to over 1.0 μm.

Therefore the glycocalyx layer might be important on interaction between blood and endothelium, and various studies have been worked on the glycocalyx properties. The glycocalyx consists of protein, glycolipid, and proteoglycans, including exposed charged groups. The membrane-bound molecules such as [7, 8] selectins and integrins, involved in immune reactions and inflammatory processes [9, 10].

The intracellular uptake of macromolecule is regulated by glycocalyx properties. The glycocalyx surface has a negative charges, because the glycocalyx has some acidic mucopolysaccharide sidechains (glycosaminoglycan (GAG)), which contain many carboxyl groups and sulfate groups. And most of plasma proteins such as albumin has a negative charge. Thus we may anticipate that the glycocalyx layer and anionic proteins have been electrostatically repulsing with anionic proteins. In previous study [11, 12], when glycocalyx or albumin was neutralized, the permeability increased than control condition.

The glycocalyx thickness may be associated with shear stress. Haldenby et al. [13] reported that the glycocalyx thickness varies dependent on the region of vessel. This report indicates that the glycocalyx thickness might be associated with shear stress, since shear stress varies with various blood vessels. In addition, Wang et al. [14] reported that the glycocalyx thickness was thin for low shear region such as a sidewall of the bifurcation, and thick for high shear region such as divider of the bifurcation.

In this study we focused on the glycocalyx thickness affected by shear stress. We measured the albumin uptake into the cultured endothelial cells with imposed shear stress stimulus along with the visualization of glycocalyx layer. To see the effect of surface charge of glycocalyx layer, we measured the albumin uptake on neutralized glycocalyx layer [15].

2   Materials and Methods

2.1   Cell culture

Cultured bovine aorta endothelial cells (BAECs; lot. 32010, Cell Systems, U.S.A.) were used in all experiments. The BAECs were seeded in 25 cm2 culture flasks (3014, Falcon, U.S.A.) and cultured in Dulbecco's modified Eagle's medium (DMEM; 31600-34, GIBCO, U.S.A.). BAECs of passages 5-9 were used for the measurement. Subsequently BAECs were seeded on 2% collagen type IV (CELLMATRIX-4-20, Nitta Gelatin, Japan) coated glass base dish (3910-035, IWAKI, Japan) or plastic bottom dish (Falcon, U.S.A.) which was used to prepare specimen for observation with electron microscopy, after reaching the confluency in 7-10 days.

2.2   Shear stress loading

The culture medium was used as perfusate. The rectangular flow chamber (height: 0.02 cm, width: 2.0 cm, length: 1.20 cm) was placed on the dish with BAECs and attached to the section A in the flow circuit. Shear stress at the wall was given by (1):

$$\tau = 6\mu Q/h^2 \cdot b \tag{1}$$

where τ: wall shear stress (Pa), μ: the viscosity ($8.5 \times 10\text{-}4$ Pa·s: measured by a rotational viscometer at 37℃), Q: flow rate (cm3/s), h: the flow channel height (0.02 cm), and b: the flow channel width (2.0 cm). The wall shear stress ranges between 0.5-3.0 Pa. The BAECs was exposed to the shear stress for 48 hours. The perfusate was kept at 37℃ by water bath and gassed with 5% CO_2-95% air to maintain the pH 7.3 throughout the experiment.

2.3   Albumin uptake

The albumin uptake into the cell is determined by measuring the fluorescence of Tetramethylrhodamin isothiocyanate conjugated albumin (TRITC-albumin; A-847, Molecular Probes, U.S.A.). After the shear stress had been loaded to the endothelium for 48hours, the flow circuit was exchanged with another flow circuit to uptake the TRITC-albumin and the BAECs was supplied the TRITC-albumin for an hour as loaded the same shear stress.

2.4   Acquisition and analysis of images

After the TRITC-albumin loading, the flow chamber was removed from the flow circuit and the DMEM containing TRITC-albumin in the flow chamber was washed away for 5 minutes with fresh medium without TRITC-albumin. The flow chamber was attached to the previous flow circuit again and the BAECs was observed as loaded the same shear stress, too.

The observation was conducted by conformal laser scanning microscopy (MRC 600 system, BIO-RAD Microscience, U.S.A.) The flow chamber was mounted on the stage of the invert microscopy, and the fluorescence images and the transmission images were acquired at five random sections respectively. Each fluorescence image was consisted of eleven serial tomographic images which were acquired for vertical direction (height direction of cell) every 1 μm. We obtained the control uptake data from the same lot and the same generation of cells without shear stress stimulus. The procedure of operation is shown in Figure 1. The acquired fluorescence images were analyzed by an image analysis software (NIH Image) with personal computer (Macintosh G4, Apple computer).

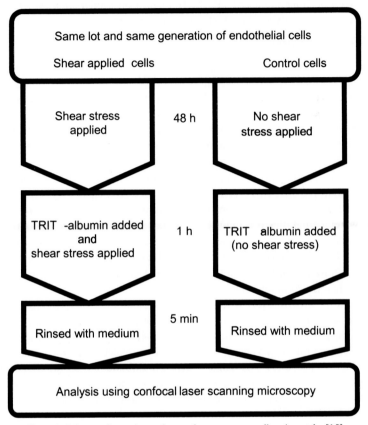

Figure 1. Schema of experimental procedure to measure albumin uptake [15].

2.5   Investigation of glycocalyx

After the shear stress loading, specimens for investigating glycocalyx layer by a scanning electron microscopy (SEM) were prepared. The perfusate was removed, and the BAECs were washed in PBS (+) (05913, Nissui, Japan) and 0.2 M sodium cacodylate buffer (29810, TAAB, U.K.). Then the BAECs were prefixed in containing 3.6% glutaraldehyde (EM Grade, 35330, TAAB, U.K.), 1500 ppm ruthenium red, 0.2 M sodium cacodylate buffer mixture fixative for 24 hours at room temperature. Then the BAECs were rinsed in 0.2 M sodium cacodylate buffer with sucrose. The BAECs were postfixed in 1.0% osmium tetroxide, 1000 ppm ruthenium red, 0.2 M sodium cacodylate buffer mixture for 3 hours at room temperature. The BAECs on the plastic dish were then dehydrated in acetone and embedded in Epon. Ultrathin sections were made after Epon layer had been peeled off the plastic in order to retain cellular orientation. Acquired images were analyzed with an image analyze soft (NIH Image) on a computer (Macintosh G4, Apple computer) and glycocalyx thickness was measured.

2.6 Cell surface charge measurement

After the shear stress loading, charge level of the glycocalyx layer on endothelial surface was decided by measuring the amount of toluidine blue (Sigma, U.S.A.) which was cationic particle and has been used to measure the charge on various cells [16].

The perfusate was removed, and the BAECs were washed in PBS (+) and 0.25 M sucrose (31365-1201, Junsei Chemical, Japan). Then the BAECs were stained in 0.25 M sucrose, 0.001% toluidine blue mixture for an hour at 4℃. We used 0.25 M sucrose, since Thethi et al. [17] reported that in presence of 0.25 M sucrose, the binding of toluidine blue was insensitive to pH between 3 and 9. Then the BAECs were washed in distilled water 5 times for 10 minutes. The BAECs were soaked in distilled water containing 0.1 mg/ml protamine sulfate (Type X, P-4020, Sigma, U.S.A.) to extract the toluidine blue, which was attaching to the glycocalyx, for 30 minutes at 4℃. The toluidine blue is displaced with the protamine sulfate since the protamine sulfate has higher affinity for carboxyl groups and sulfate groups in the glycocalyx than the toluidine blue.

The extract was transferred to quartz glass cuvettes. Then the absorbance of the extract was measured by using a double beam spectrophotometer (U-3400, Hitachi, Japan) at the extinction wavelength of 640 nm. The absorbance indicates amount of charge on the glycocalyx since the absorbance is a concentration of the toluidine blue, which electrostatically attaches to anionic site on the glycocalyx, in the extract. From the measured absobance, the average absorbance per each cell was calculated. The cell amounts on the each dish were measured by using a countering chamber.

2.7 Neutralization of glycocalyx charge

The glycocalyx was neutralized using protamine sulfate (Type X, P-4020, Sigma, U.S.A.). Protamine sulfate is small and charges highly positive at physiological pH, since its isometric point (pI) is 10-12. Therefore, protamine sulfate has been used as such charge neutralizer [11]. Protamine sulfate was added to perfusate for a final concentration of 0.001 mg/ml.

After the shear stress had been loaded to the endothelium for 48hours, the flow circuit was exchanged with another flow circuit to neutralize the glycocalyx and the BAECs was supplied protamine sulfate for 30 minutes as loaded the same shear stress. After protamine sulfate loading, the flow chamber was removed from the flow circuit and the DMEM containing protamine sulfate in the flow chamber was washed away with fresh medium. Then the BAECs were loaded TRITC-albumin and measured the light intensity as stated above. In addition, as the referential experiments, the non-treatment and protamine sulfate treated control BAECs were similarly measured.

3 Results

3.1 Albumin uptake

On the cells, which were loaded shear stress at 1.0 Pa, large albumin uptake was occurred compared with control (no-flow). At 3.0 Pa, on the other hand, albumin uptake was lesser than that of control. At 0.5 and 2.0 Pa, there was not significant difference between shear stress loaded endothelial cell and control cell.

We quantitatively evaluated the relation between the shear stress and the albumin uptake (Fig. 2). A vertical axis indicates the relative light intensity on shear stress loaded cells compared with average light intensity on control. And the light intensity indicates the alubmin uptake of endothelial cells. There was no significant difference between 0.5Pa shear stress loaded cells and control cells. However, at 1.0 Pa, albumin uptake increased 16.4% of control and there was significant difference (P < 0.05). And at 3.0 Pa, the albumin uptake was decreased to 26.7% of control (P < 0.001).

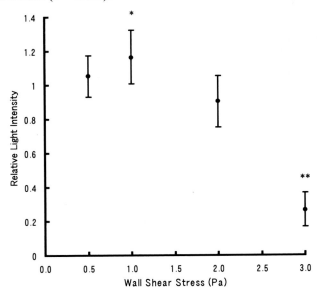

Figure 2. Effect of shear stress on albumin uptake. The relative light intensity indicates the ratio of the uptake with applies shear stress to that without [15].

3.2   *Investigation of glycocalyx*

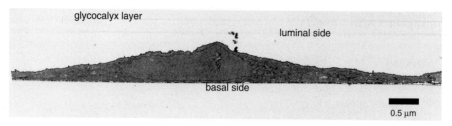

Figure 3. TEM image of the cross-section of an endothelial cell covered with glycocalyx [15].

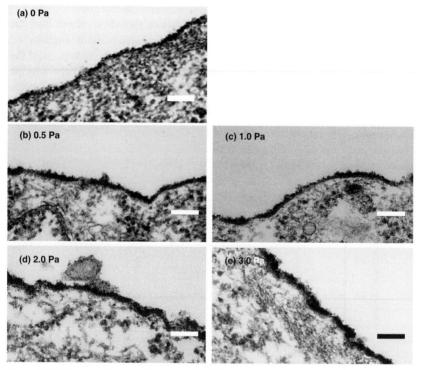

Figure 4. TEM images of clycocalyx layers (a) without shear stress and with sheare stress of (b) 0.5 Pa, (c)1.0 Pa, (d) 2.0 Pa, and (e) 3.0 Pa. Magnification was 20,000 [15].

Fig. 3-5 shows glycocalyx layer images, which were acquired by scanning electron microscopy. The black layer on the endothelial surface is the glycocalyx layer (Fig. 3), which was stained by ruthenium red. The endothelial surface was evenly covered with the glycocalyx. And the gap junction was also covered with the glycocalyx as well as the endothelial surface. In addition, The glycocalyx covered over areas where vesicles were going to detach from the endothelial membrane. We exhibited glycocalyx images on control cells and shear stress loaded cells in Fig. 4a-e. To outward seeming, all the shear stress loaded endothelial glycocalyx

(a) SEM image

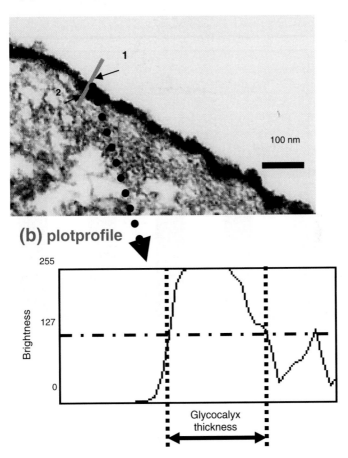

Figure 5. Determination of the thickness of the glycocalyx layre [15].

layer surfaces were ruder than that of control. But, there wasn't different of amplitude of rough on glycocalyx layer surface among each shear stress. And there wasn't different of density of stained ruthenium red among each condition, either. However, glycocalyx layer thickness was different from some conditions to some extent. Then, we measured the glycocalyx thickness from each image by following manner. First, we acquired electron microscopic images with a 256 gray scale with NIH image and draw a vertical line across the glycocalyx layer against endothelial membrane. Second, we measured brightness on the line and drew a plotprofile of the brightness. Third, we measured the glycocalyx thickness (Fig. 5). The glycocalyx thickness was defined as a length between a middle of up slope to go up

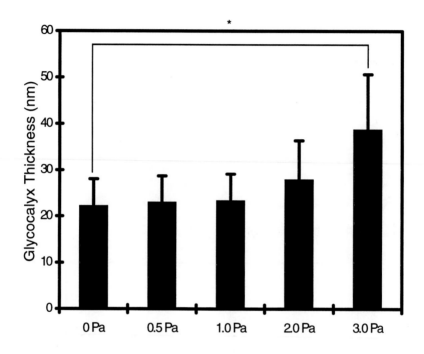

Figure 6. Effect of applied shear stress on glycocalyx.

the highest brightness and a middle of down slope from the highest brightness in plotprofile. Because electron microscopy shows higher brightness as the atomic weight of detected material becomes high, the brightness of background was zero and the glycocalyx labeled with ruthenium red was reflected darker than others. The glycocalyx thickness was measured at 30 point per cell. The measured glycocalyx thickness on control, loaded shear stress (0.5, 1.0, 2.0, 3.0 Pa) were appeared in Fig. 6. There was no significant difference between control and 0.5, 1.0, 2.0 Pa, but 3.0 Pa was significantly different from control ($P < 0.05$). The glycocalyx thickness of 3.0 Pa was about 74% thicker than control.

3.3 Shear-induced change on endothelial surface charge

We showed change on the endothelial surface charge after the shear stress loading for 48 hour and control in Fig. 7. A vertical axis indicates relative absorbencies, which measured absorbencies were divided by measured cell number with a counting chamber. Because there is a proportional relation between absorbency and charge content of glycocalyx [16], We estimated relative absorbency as charge content. As shown in Fig. 7, there was no significant difference between control and 0.5, 1.0, 2.0 Pa like the glycocalyx thickness. But there was significant

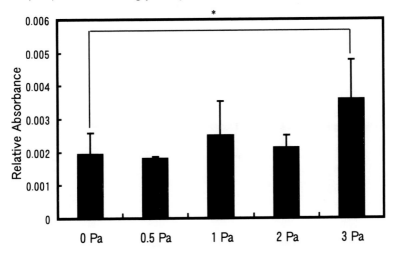

Figure 7. Effect of shear stress on surface charge. The relative absorbance indecates the surface charge of an individual cell and was defined as the measured absorbance divided by the number of cells [15].

difference between control and 3.0 Pa (P < 0.05), and the glycocalyx thickness of 3.0 Pa was about 84% thicker than control.

3.4 Effect of neutralized charge on shear-dependent albumin uptake

In Fig. 8, we showed the relation between the shear stress and the albumin uptake on neutralized glycocalyx charge with protamine sulfate and non-treatment. A vertical axis indicates the relative light intensity of both neutralized charge and non-treatment versus the average light intensity on the control cells. The black plots indicate the change of albumin uptake with neutralized charge, and the white plots indicate the change of albumin uptake with non-treatment. Although there was no significant difference between neutralized charge and non-treatment at 2.0 Pa, there was significant difference between neutralized charge and non-treatment at 0.5 Pa (P < 0.05), 1.0 Pa (P < 0.05), 3.0 Pa (P < 0.001). At 0.5 Pa and 1.0 Pa, albumin uptake of neutralized charge increased about 20% of non-treatment, and at

3.0 Pa increased about 70% of non-treatment. At 0.5 Pa and 1.0 Pa, albumin uptake increased about 20% of non-treatment. The increment indicates baseline charge, which is a non-treatment control charge, since this profile indicates apparent increment on neutralized charge including charge according to baseline charge.

4 Discussion

4.1 Albumin uptake

This experiment demonstrated that the albumin uptake changed dependent on the shear stress (Fig. 2). The albumin uptake tended to increase up to 1.0 Pa, and

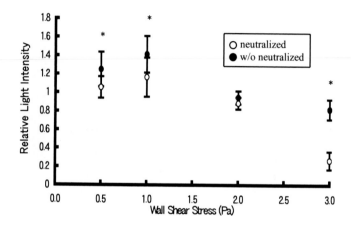

Figure 8. Effect of shear stress on albumin uptake with and without the charge of glycocalyx being neutralized.

decrease on over 1.0 Pa. This result supports the low shear hypothesis on development points of atherosclerosis, which Caro et al. [1] invoked. Because there are some reports that depositions of large molecular proteins, such as low density lipoprotein (LDL) [2], albumin [3, 4], and horse radish peroxidase (HRP) [14, 18], were found in initial symptom of atherosclerosis. In addition, we compared this result with Kudo's study [5]. There were some differences on the magnitude of albumin uptake, but our data agreed with Kudo's data in tendency that albumin uptake was elevated at low shear stress (0-1.0 Pa), and suppressed at high shear stress (over 1.0 Pa). We used bovine aorta endothelial cells, while Kudo used porcine aorta endothelial cells. Therefore, this tendency doesn't depend on species. In the Kudo's study [5], the TRITC-albumin was administered to the endothelium after stopped loading the flow to conserve the constant albumin diffusion. On the other hand, in our study, the TRITC-albumin was administered to the endothelium while the endothelium was being loaded the flow to approach to intravital condition. Despise of such difference on manner of administration, the tendency on our study corresponded with Kudo's study. This correspondence indicates that the effect of endothelial function modified by shear stress may be more efficient than the effect of flow velocity on albumin uptake.

However, in this experiment, at high shear stress (When main stream velocity is faster.), albumin uptake tended to decrease. Moreover, this tendency corresponded with Kudo's study [5], in which TRITC-albumin was loaded after flow was stopped. In our experiment, TRITC-albumin was loaded as the perfusate was flowing. Therefore, according to this experiment, We suggest that on albumin uptake, the effect of stream velocity is small, and the endothelial function, which controls albumin uptake by being subjected to shear stress, is more efficient.

4.2 Change of glycocalyx layer thickness

In this study, we demonstrated that in vitro the glycocalyx layer labeled with ruthenium red covered almost all over the endothelial surface membrane (Fig. 3-5). In previous study [6] which we imitated the stain and the fixative manner, the glycocalyx layer had thickness in range of 20 nm. According to our data, the glycocalyx thickness range was 20-40 nm. Therefore, we suggest that the glycocalyx thickness is adequate. The glycocalyx layer was irregularly shaped with a fluffy appearance.

This study also demonstrated that the glycocalyx thickness increased at high shear stress (3.0 Pa), while did not change against control at low shear stress (0.0-2.0 Pa) (Fig. 5). This result indicates that the glycocalyx thickness changes depend on shear stress over 3.0 Pa. Wang et al. [14] reported that in rabbit aorta bifurcation the glycocalyx thickness was thicker at high shear stress region than at low shear stress region. Therefore, our results corresponded with Wang's report. Baldwin et al. [19] investigated glycocalyx layers at rabbit aortic endothelium with a ruthenium red and an electron microscopy and reported that the glycocalyx thickness was about 20 nm. And Rostgaard et al. [20] investigated glycocalyx layers at rat capillary in intestinal villus with a ruthenium red and an electron microscopy and reported that the glycocalyx thickness was about 50 nm. These results indicate that glycocalyx thickness may be thin at low shear stress and thick at high shear stress, since shear stress differs according to various vessels. In addition, our result indicates that shear stress induces the change of endothelial surface charge. The increment of glycocalyx layer labeled with cationic dye, ruthenium red, means the increment of charge. But, it is not clear whether the glycocalyx change is caused by increasing the number or length of glycosaminoglycan (GAG) on a coreprotein, or by enlarging the length of coreprotein. Arisaka et al. [21] reported about effects of shear stress on GAG and protein metabolism of porcine aorta endothelial cells in vitro, and demonstrated that the application of shear stress over 24 hours led to significantly increased both GAG and protein synthesis. Therefore, enlarging proteoglycan, which consists of coreprotein and GAG, may cause the increment of glycocalyx thickness.

4.3   Change of surface charge

This experiment demonstrated that charge content on an endothelial surface did not change at comparatively low shear stress (0.5, 1.0, 2.0 Pa), and increased at comparatively high shear stress (3.0 Pa) against control (Fig.7). Comparing with the change of glycocalyx thickness, both changes on charge content and glycocalyx thickness had a little difference at comparatively low shear stress. The change on glycocalyx thickness was about 74% increment against control, while the change on charge content was about 84% increment against control. Although this 10% difference may be caused with a change on distribution of glycocalyx on an endothelial cell, this data indicate that change on charge content is mainly induced by change on glycocalyx thickness.

In this experiment, although charge was measured by Van Damme's method [22], there was a problem on staining glycocalyx. Toluidine blue, which was used to electrostatically label the glycocalyx, was absorbed into intracellular space. In this study, glycocalyx was stained at 4°C to stop vesicle mediated metabolic transport. Kudo et al. [23] reported that metabolic transport was inhibited at 4°C. Since molecular weight of toluidine blue is 308 Da, we can regard toluidine blue transport from extracellular space to intracellular space as free diffusion. And we can regard Endothelial surface area as constant surface area on all conditions according to our data. Therefore, we regarded amount of absorbed toluidine blue into intracellular space as constant on all conditions and judged that this method was adequate on comparison of relative change.

4.4   Effect of neutralized charge on albumin uptake

In this experiment, we demonstrated that neutralized glycocalyx charge by protamine sulfate did not effect to albumin uptake at comparatively low shear stress (0.5, 1.0, 2.0 Pa), and at comparative high shear stress (3.0 Pa), albumin uptake on neutralized charge increased about two fold of albumin uptake on non-treatment. This result indicates that the change of glycocalyx at comparative high shear stress (3.0 Pa) caused decay on albumin uptake dependent on shear stress, and at low shear stress (0.5, 1.0, 2.0 Pa), doesn't effect shear-stress-induced change on albumin.

In this experiment, although, the change of albumin uptake against non-neutralized charge was caused by neutralized glycocalyx charge at 3.0 Pa, shear-stress-depend change on albumin uptake appeared despite of neutralized glycocalyx charge. It isn't enable to explain this shear-stress-dependent change on albumin uptake without the effect on charge by our study. However, this residual change may connect to Kudo's study [24]. In the light of albumin uptake caused by energy-dependent transport by transcytosis, Kudo et al. [24] studied about effect of shear stress on ATP-dependent albumin transport. According to the result, they indicated that ATP and albumin uptake increased at low shear stress (1.0 Pa), and ATP and albumin uptake decreased at high shear stress (6.0 Pa). Thereby, the residual change on albumin uptake in this study may have relation to energy-dependent transport.

In this study, we used protamine sulfate as an anionic charge neutralizer. But, it is possible that increment on albumin uptake may be due to a cytotoxic effect on endothelium as well as protamine sulfate's neutralization of charge barrier on the endothelial surface. However, we confirmed that protamine sulfate didn't have a cytotoxic effect. Because the viability of endothelium, which was administered with protamine sulfate, was $97.1 \pm 1.8\%$ according to dye exclusion test with 1.0% trypan blue. In addition, Swanson et al. [11] reported that no mediator for protamine-induced increased microvascular albumin permeability was identified. Thereby, we regarded protamine-sulfate-induced increment on albumin uptake as caused by neutralized the glycocalyx charge.

Acknowledgements

This work was supported by Grant-in-Aid for Scientific Research on Priority Areas 15086214 from the Ministry of Education, Culture, Sports, Science and Technology of Japan.

References

1. Caro, C. G., Fitz-Gerald, J. M., Schroter, M. C., 1969. Arterial wall shear and distribution of early atheroma in man. Nature 233, 1159-61.
2. Schwenke, D. C., Carew, T. E., 1989. Initiation of atherosclerotic lesions in cholesterl. -fed rabbits-Forcal increases in arterial LDL concentration precede development of fatty streak lesions-, Arteriosclerosis 9 (6), 895-907.
3. Packham, M. A., Roswell, H. C., Jorgensen, L., Mustard, J. F., 1967. Localized protein accumulation in the wall of the aorta., Exptl. Molec. Pathol. 7, 214-32.
4. Somer, J. B., Schwartz, J., Focal, 1972. 3H-cholesterol uptake in the pig aorta. 2. Distribution of 3H-cholesterol across the aortic wall in areas of high and low uptake in vivo. Arteriosclerosis 16 (3), 377-88.
5. Kudo, S., Ikezawa, K., Matsumura, S., Ikeda, M., Oka, K., Tanishita, K., 1996. Effect of shear stress on albumin uptake into cultured endothelial cells. ASME Adv. Bioeng. BED-33, 209-10.
6. Luft, J.H., 1966. Fine structure of capillary and endocapillary layer as revealed by ruthenium red. Microcirc. Symp. Fed. Proc. 25, 1773-2783.
7. Risau, W., 1995. Differentiation of endothelium. FASEB J. 9, 926-33.

8. Siegel, G., Malmsten, M., 1997. The role of the endothelium in inflammation and tumor metastasis. Int. J. Microcirc. Clin. Exp. 17, 257-72.

9. Ley, K., 1996. Molecular mechanisms of leukocyte recruitment in the inflammatory process. Cardiovasc. Res. 32, 733-42.

10. Springer, T.A., 1995. Traffic signals on endothelium for lymphocyte recirculation and leukocyte emigration. Annu. Rev. Physiol. 57, 827-72.

11. Swanson, JA., Kern, D. F., 1994. Characterization of pulmonary endothelial charge barrier. Am. J. Physiol. 266, H1300-3.

12. Vink, H., Duling, B. R., 2000. Capillary endothelial surface layer selectively reduces plasma solute distribution volume. Am. J. Physiol. 278, H285-9.

13. Haldenby, K.A., Chappell, D. C., Winlove, C. P., Parker, K. H., Firth, J. A., 1994. Focal and regional variations in the composition of the glycocalyx of large vessel endothelium. J. Vasc. Res. 31, 2-9.

14. Wang, S., Okano, M., Yoshida, Y., 1991. Ultrastructure of endothelial cells and lipid deposition on the flow dividers of brachiocephalic and left subclavian arterial bifurcations of the rabbit aorta. Doumyakukouka 19 (12), 1089-100.

15. Ueda, A., Shimomura, M., Ikeda, M., Yamaguchi, R., Tanishita, K., 2004. Effect of glycocalyx on shear-dependent albumin uptake in endothelial cells, American J. Physiology (in press)

16. Van Damme, M. P. I., Tiglias, J., Nemat, N., Preston, B.N., 1994. Determination of the charge content at the surface of cells using a colloid titration technique. Anal. Biochem. 223, 62-70.

17. Thethi, K., Jurasz, P., MacDnald, A. J., Befus, A. D., Man, S. F. P., Duszyk, M., 1997. Determination of cell surface charge by photomeric titration. J. Biochem. Biophy. Methods 34, 137-45.

18. Yoshida, Y., Sue, W., Okano, M., Oyama, T., Yamane, T., Mitsumata, M., 1990. The effect of augmented hemodynamic forces on the progression and topography of atherosclerotic plaques. Ann. N. Y. Acad. Sci. 598, 256-73.

19. Ono, N., 1995. A Semi-quantitative measurement of glycocalyx and an ATP bioluminescent assay for the analysis of pseudomonas aeruginosa biofilm., Nippon Hinyokika Gakkai Zasshi 86 (9), 1440-9.

20. Caro, C.G., Fitz-Gerald, J.M., Schroter, R.C., 1971. Atheroma and aterial wall shear observation, correlation and proposal of a shear dependent mass transfer mechanism for atherogenesis. Proc. Roy. Soc. Lond. B. 177, 109-59.

21. Baldwin, A. L., Winlove, C. P., 1984. Effects of perfusate composition on binding of ruthenium red and gold colloid to glycocalyx of rabbit aortic endothelium. J. Histochem. Cytochem. 32 (3), 259-66.

22. Rostgaard, J., Qvortrup, K., 1997. Electron microscopic demonstrations of filamentous molecular sieve plugs in capillary fenestrate. Microvas. Res. 53, 1-13.

23. Kudo, S., Ikezawa, K., Matsumura, S., Ikeda, M., Oka, K., Tanishita, K. Relationship between energy-dependent macromolecule uptake and transport

granules in the endothelial cells affected by wall shear stress. Transactions of the Japan Society of Mechanical Engineers 623 (64) B, 2123-31.

24. Kudo, S., Morigaki, R., Saito, J., Ikeda, M., Oka, K., Tanishita, K., 2000. Shear-stress effect on mitochondrial menbrane potential and albumin uptake in cultured endothelial cells. Biochem. Biophys. Res. Commun. 270, 616-21.

BIOMECHANICAL AND BIOTRIBOLOGICAL IMPORTANCE OF SURFACE AND SURFACE ZONE IN ARTICULAR CARTILAGE

T. MURAKAMI, Y. SAWAE, N. SAKAI AND I. ISHIKAWA

Department of Intelligent Machinery and Systems, Kyushu University,
Hakozaki, Higashi-ku, Fukuoka 812-8581, Japan
E-mail:tmura@mech.kyushu-u.ac.jp

The natural synovial joints have excellent tribological performance known as very low friction and very low wear for various daily activities in human life. These functions are likely to be supported by the adaptive multimode lubrication mechanism, in which the various lubrication modes appear to operate to protect articular cartilage and reduce friction, depending on the severity of the rubbing conditions. In this paper, the biomechanical and biotribological importance of surface and surface zone in articular cartilage is described in relation to frictional behavior and deformation. It is pointed out that the replenishment of gel film removed during severe rubbing is likely to be controlled by supply of proteoglycan from the extracellular matrix, where the chondrocyte plays the main role in the metabolism. The roles of surface profiles and elastic property measured by atomic force microscopy are described. It is shown that the local deformation of biphasic articular cartilage remarkably changes with time under constant total compressive deformation. The importance of clarification of actual strain around chondrocyte is discussed in relation to the restoration of damaged articular cartilage.

1 Introduction

The natural synovial joints with very low friction and low wear are likely to operate in the adaptive multimode lubrication mechanism [1-3], in which the fluid film lubrication, weeping, adsorbed film and/or gel film lubrication can become effective depending on the severity of operating conditions in various daily activities. In natural hip and knee joints during walking, the elastohydrodynamic lubrication (EHL) mechanism, based on the macro- and micro-scopic elastic deformation of articular cartilage and viscous effect of synovial fluid, appears to play the main role in preserving low friction and low wear. The effectiveness of the fluid film lubrication depends on the minimum film thickness, which should be higher than the surface roughness. The maximum height of the undeformed articular cartilage surface ranges from 1 to 2 μm [4]. Fortunately, even in the contact conjunction zone (load-carrying zone) where the minimum film thickness is less than 1 μm, the flattening of initial protuberances of articular cartilage surface is enable to achieve fluid film lubrication during walking, as indicated by micro-EHL analyses [5, 6]. At start-up after long standing, however, some local direct contact may occur even in healthy natural joints. Therefore, the protecting performance of the surface film on articular cartilage becomes important in the mixed or boundary lubrication regime.

In the rubbing condition with local contact, the adsorbed films [3, 7-10] such as phospholipds, glycoproteins and proteins appear to protect the rubbinfg surfaces and reduce wear. However, the severer rubbing is likely to remove the adsorbed film and rub off the underlying surface layer. For this condition, natural synovial joints seem to have another protective layer of gel film with low shearing resistance.

The repair of damaged cartilage depends upon not only the replenishment of the surface layer but also the bulk matrix recovery. It is considered that articular cartilage adapts to changing mechanical environments where the chondrocyte can respond to the local stress-strain state [11, 12], but the detailed process has not yet been clarified. Chondrocytes are concerned with the mechanisms underlying remodeling, adaptation, and degeneration of articular cartilage in joints subjected to changing loads, and it is important to know the stress-strain state of the cartilage and extracellular matrix around chondrocytes. Its depth-dependent complicated structure causes a complex stress-strain field. Some compression studies [13, 14] using confocal laser scanning microscopy (CLSM) or video microscopy focused on the equilibrium strain state of a compressed articular cartilage specimen. However, articular cartilage has a viscoelasticitic property based on a high water content of up to 80% and the fluid-flow behavior concerns the time-dependent deformation of articular cartilage when the tissue is exposed to compression. Therefore, it is necessary to observe the time-dependent stress and strain of compressed articular cartilage, to clarify the change of the mechanical environment of chondrocytes.

In this paper, first the protective mechanism provided by gel films is shortly described. Next, we examine cartilage surface properties such as surface profile and elasticity. Then, time-dependent and depth-dependent local strain in articular cartilage under constant total compression is observed. Finally we discuss the importance of actual strain around chondrocyte particularly in surface zone in relation to the restoration of damaged articular cartilage.

2 Protective Roles of Surface Layer of Articular Cartilage

To elucidate the protective mechanism of the adsorbed and gel films, the reciprocating tests [15, 16] of ellipsoidal specimen of articular cartilage from porcine femoral condyle against a glass plate were conducted at a constant load of 4.9 N (mean contact pressure : 0.13 MPa), stroke of 35 mm, and sliding speed of 5 mm/s, under both unlubricated and lubricated (with saline) conditions as shown in Fig.1. The rubbing tests were stopped at definite sliding distances. To investigate the changes in surface morphology with repetition of rubbing, the articular cartilage surfaces were observed in saline by atomic force microscope (AFM) as fluid tapping mode using scanning probe microscopy SPM systems (Nanoscope III, Digital Instruments, USA). By this noncontact method immersed in liquid, the influence of capillary force between the tip and the observed surface was diminished and the rubbing action by the tip was minimized. The scanning speed is 1 to 2 lines/s. Figure 1 shows the changes in the frictional behaviors in sliding pairs of a natural articular cartilage specimen and glass plate with the sliding distances under both unlubricated and saline lubricated conditions. For both conditions the friction was increased with the sliding distance. However, under an unlubricated rather than a lubricated condition, a lower friction was maintained, thus the role of the surface layer of the lubricating adsorbed films and/or gel films was expected to be emphasized. Under lubricated conditions with low viscosity saline, the easy removal of lubricating molecules or surface layers from the cartilage surface seems to be responsible for a rise in friction. The preserving of low friction is ascribed to the surviving of adsorbed molecules and/or gel film in the case of longer rubbing.

To examine the friction mechanism in the transient rubbing process for the unlubricated condition, the articular surfaces at 0.35 m and 9 m (Fig.1) were observed by the fluid tapping mode AFM. Figure 2(a) shows the AFM images of an intact articular cartilage surface, which has a considerably smooth morphology with a maximum height of 1 to 2 μm. At a short sliding distance of 0.35 m under the unlubricated condition where the coefficient of friction is less than 0.01, the cartilage surface was slightly rubbed as shown in Fig.2(b). With further rubbing, the fibrous tissues appeared at 9 m sliding in Fig.2(c), where coefficient of friction becomes about 0.1. Therefore, it is considered that the acellular and non-fibrous surface layer has been rubbed off at this stage. As the detailed structure of gel

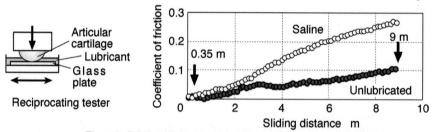

Figure 1. Frictional behaviors of articular cartilage against glass plate.

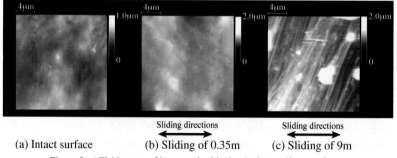

(a) Intact surface   (b) Sliding of 0.35m   (c) Sliding of 9m

Figure 2. AFM images of intact and rubbed articular cartilage surfaces.

film has not yet been clarified, we conducted the treatment of intact articular cartilage with Chondroitinase ABC as the enzyme specific for proteoglycans [17]. This treatment for 30 h denuded its amorphous layer and exposed the underlying fibrillar network. Therefore, the main constituent of the most superficial gel layer consisting of acellular and nonfibrous tissue was identified as proteoglycans. This gel film layer is considered to play a protective role like a kind of solid-lubricant or as a surface gel hydrated lubricant [18] after removal of adsorbed films in the thin film mixed and boundary lubrication regimes.

On the basis of these results, the local direct contact point after removal of the surface films composed of adsorbed and gel films is depicted in Fig.3. The replenishment of surface films is likely to be brought about by the supply of amphiphilic molecules and hyaluronic acid in synovial fluids and the proteoglycans from the extracellular matrix in which the chondrocyte can control the metabolism.

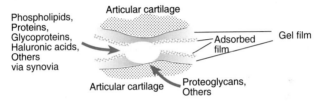

Figure 3. Replenishment of damaged surface films of articular cartilage via synovial fluid and extracellular matrix.

3 Surface Profile and Elastic Property of Cartilage Surface

Superficial articular cartilage is covered with adsorbed film and acellular amorphous layer of 200-500 nm thick which appears to play an important role in the tribological function of synovial joints [15]. In this study, we used AFM to investigate this surface layer. The surface topography of articular cartilage was observed, and the surface stiffness was characterized by measuring force-indentation curves. Such microscopic properties are necessary to know adequately effects of mechanical environment on the metabolism of articular cartilage.

Porcine knee joints were obtained from a local butcher shop and stored at 4 °C until experiment. Rectangular cartilage plates (approximately 2 × 2 × 0.5 mm) were separated from the surface of femoral condyles as shown in Fig.4 and immediately glued onto a circular glass coverslip using cyanoacrylate glue. Using two-sided adhesive tape, the coverslip was fixed to a steel disc mounted on the piezoscanner of the AFM. A Nanoscope IIIa AFM was used to image the surface topography of articular cartilage. All experiments were conducted in an aqueous environment using the liquid cell of the AFM. PBS was used as hydration media.

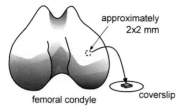

Figure 4. Cartilage specimen.

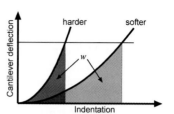

Figure 5. Relative elasticity by FIEL.

In addition to the observation of surface topography, surface elasticity was measured by AFM [19]. Using AFM, elasticity measurements are performed by pushing a tip of AFM onto the surface and deriving force-versus-indentation (F-I) curves. Here, cantilevers with a nominal force constant of $k = 0.58$N/m and oxide-sharpened Si_3N_4 tips were used. To calculate relative elasticity values, we used the FIEL (force integration to equal limit) equations described by Hassan et al. [20]. These equations are given for conical, parabolic and cylindrical tips as follows;

$$E^*_1 / E^*_2 = (w_1 / w_2)^{-n} \tag{1}$$

where E^* is equivalent modulus of elasticity, and w is the area under a F-I curve (Fig. 5), $n = 2$ for cone, 3/2 for parabola, 1 for cylinder. E^* is defined by

$$\frac{1}{E^*} = \frac{1-v_{tip}^2}{E_{tip}} + \frac{1-v_{sample}^2}{E_{sample}} \approx \frac{1-v_{sample}^2}{E_{sample}} \tag{2}$$

where E and v are Young's modulus and Poisson's ratio, respectively.

The surface of intact porcine articular cartilage was very smooth, with gentle mounds of 1-2μm height under the physiological wet condition (Fig. 6). Figure 7 shows [Cantilever deflection]-[Separation] curves that were collected on 9 points

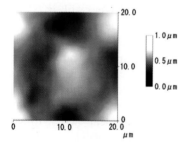

Figure 6. AFM image of porcine articular cartilage surface.

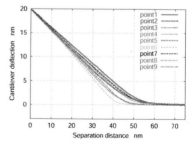

Figure 7. Cantilever deflection - Separation curves of porcine articular cartilage.

that were central point (point 5) and surrounding ones on lattice with 10 μm mesh on a surface of porcine articular cartilage in Fig.6. These curves did not vary widely, thus the surface stiffness of these points was not different significantly each other.

Surface morphology of normal natural articular cartilage under AFM showed a smooth surface. This result is consistent with previous studies. However, wrinkles, hollows or pits existed on human osteoarthritic cartilage in previous study [19].

The surface stiffness of normal porcine articular cartilage was approximately homogeneous and the average value of E^* was estimated as 2.6 $\pm$0.21 (SD) MPa, where $E^* = 2.8$ MPa for silicone rubber was used as reference. The indentation depth was about 20-45 nm, and F-I curves were integrated from 0 nm to 20 nm deflection to compare surface stiffness relatively. In our previous measurement, elastic modulus E for porcine cartilage bulk specimen at equilibrium was 2.0 $\pm$0.7 (SD) MPa [21]. Although the estimation of E value for surface depends on Poisson's ratio, intact normal cartilage surface appears to have similar elastic modulus to bulk property. In contrast, the variation of surface stiffness was found depending on surface morphology on human osteoarthritic cartilage [19]. The heterogeneity of surface stiffness should be considered for evaluation of degenerative progress of osteoarthritis.

4 Visualization of Time-Dependent Strain in Articular Cartilage in Compression

Next, we observed the changes in local strain in articular cartilage specimens under compression by monitoring the position of stained chondrocyte in the CLSM [21]. The staining of chondrocyte is treated with calcein-AM, at 1 μl/ml and 37 ℃ for 30 min. The compression apparatus shown in Fig. 8 with high precision within 0.2 μm for position control was newly developed. This apparatus was allocated on the stage of CLSM and the compression speed can be adjusted from 1 μm/s to 4,000 μm/s by feed-back control of DC servo-motor. In these tests, 13% total strain was applied in 1 s in unconfined compressive condition. On the basis of these visualized images, the time-dependent and depth-dependent changes in local strain of articular cartilage were evaluated. The fluorescence images of chondrocytes in articular caritilege specimens are shown in Fig.9, where the highlighted elliptical bodies of 10 to 20 μm indicate the chondrocytes. Some chondrocytes are enclosed with white circles. These images were observed every 2 s until 400s at equilibrium after loading. In these tests, the articular cartilage surface was located at contact with the fixed plate, in order to keep the surface position within visual field of microscope. We estimated the local strain by calculating the changes of the distance in perpendicular direction to the cartilage surface between the definite chondrocytes as follows;

Local strain immediately after compression : $(a - b)/a$

Local strain at equilibrium : $(a - c)/a$

48

Change in local strain to equilibrium immediately after compression : $(b - c) / a$
Here, as shown in Fig.9, a, b and c means the distance between the corresponding chondrocytes before compression, immediately after compression and at equilibrium after 400s, respectively.

In Fig. 10, the estimated values of local strain immediately after the total deflection of 13% and at the equilibrium are plotted against the relative location in the depth direction, where 0 means surface and 1 means the tidemark as the boundary of subchondral bone. The articular cartilage is usually discriminated as three zones, i.e., the surface, middle and deep zone along the perpendicular direction from surface. It is noted that the response of middle and deep zone is quick but the deformation of surface zone is low immediately after compression. However, the surface zone was largely compressed than average strain during stress relaxation. In contrast, the deformation of deeper zone was clearly recovered probably accompanied with flowing of fluid into the middle and deep zone during stress relaxation.

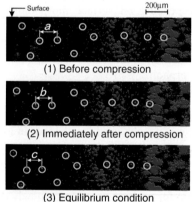

(1) Before compression

(2) Immediately after compression

(3) Equilibrium condition

Figure 9. Fluorescence images.

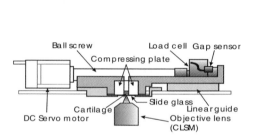

Figure 8. Compression apparatus.

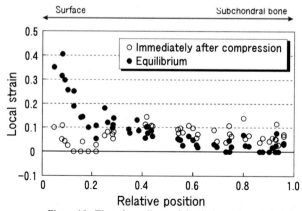

Figure 10. Time-depending and depth-depending strain behavior.

5   Discussion

As described above, the natural synovial joints are likely to operate in the adaptive multimode lubrication mechanism, in which various lubrication modes can become effective depending on the severity of operating conditions in various daily activities.   In this paper, we focus the importance of surface and surface zone in articular cartilage.   On the EHL mechanism, the fluid film formation is enhanced by elastic deformation of articular cartilage, and the possibility of fluid film lubrication mode depends on relative value of film thickness and surface roughness of cartilage.   As shown in section 3, cartilage surface has surface roughness of 1 to 2 μm height and similar elastic modulus to bulk property.   The micro-EHL analysis by Dowson and Jin [5, 6] used the surface asperity model of maximum height of 2 μm and elastic modulus of 16 MPa.   Although some modifications for pitch from peak to peak in surface topography and for larger elastic deformation with low elastic modulus are required, the similar flattening of surface asperity is expected and thus fluid film lubrication appears to be capable of protect the rubbing cartilage surfaces during normal walking.   Some degraded cartilage surface in osteoarthritic cartilage may have very lower elastic modulus in local position [19], thus excessive strain may induce further degradation.   The detailed investigation of surface morphology and elastic property is required to elucidate the progressive mechanism of osteoarthritis.

Another importance to prevent the degradation of articular cartilage is the preservation of adsorbed film and underlying gel film as indicated in section 2. Particularly, the restoration of damaged proteoglycan gel film appears to depend on the supply of proteoglycan from the extracellular matrix, in which chondrocyte can control the metabolism.   In articular cartilage, chondrocyte responds to mechanical stimuli.   Therefore, to clarify the metabolism of cartilage tissue, it is required to understand the stress-strain state in cartilage and around the chondrocyte.

The visualization method to estimate the time-dependent and depth-dependent local strain behavior in cartilage described in section 4 is a useful approach to clarify actual strain condition.   In the visualization tests of chondrocytes, we found that the local strain shows time-dependent and depth-dependent behaviors.

The experimental results indicated that the surface zone moderately responded at the initial stage of compression and then was largely compressed during stress relaxation.   In the surface zone, collagen fibrils are arranged in parallel sheets to the articular surface.   This anisotropy accompanied with less proteoglycan appears to affect the compressive strain in the surface zone.   On the contrary, the middle and deep zones containing larger percentage of proteoglycan than surface zone were quickly compressed at the initial stage and then recovered probably with water flowing in under unconfined compression.   The excessive compressive strain in surface zone at equiriblium may be related to unconfined test condition.

The strains of articular cartilage control the deformation of chondrocytes as indicated by previous experimental and numerical biphasic studies [13, 22-24]. Therefore, the strain behaviors of articular cartilage are expected to have an important influence on the biosyntheses of proteoglycans by chondrocytes with mechanical stimuli, thus proteoglycan appears to control the bulk stiffness, permeability and the restoration of the lubricating gel film on the cartilage surface.

The time-depending and depth-depending changes in local strain around chondrocyte should be clarified by comparison with numerical analyses based on biphasic theory considering non-linearity and complex property of articular cartilage at the next stage. The threshold strain values to which chondrocyte can respond in the mechanical signal transduction pathway, should be evaluated by considering the time-depending and depth-depending mechanical behaviors in biphasic cartilage. The adsorbed film and gel film seem to control the permeability through surface. Therefore, further research is required to clarify the stress-strain state in articular cartilage under various surface conditions.

6   Conclusions

The biomechanical and biotribological importance of surface and surface zone of articular cartilage is summarized from the reciprocating rubbing tests of articular cartilage, AFM measurement of surface topography and stiffness, and visualization of local strain in compressed articular cartilage.

The importance of articular cartilage surface film was indicated by the protective role of proteoglycan gel film in rubbing tests. The influence of surface roughness and elastic property of surface evaluated with AFM on fluid film lubrication was described on micro-EHL mechanism and contact phenomena.

In the visualization tests of stained chondrocytes, the time-dependent and depth-dependent behaviors in the local strain of articular cartilage could be evaluated. The optical observation in unconfined compression test indicated that the surface zone moderately responded at the initial stage of compression and then was largely compressed during stress relaxation.

Acknowledgements

Financial supports were given by Grant-in-Aid for Scientific Research on Priority Areas 15086212 from the Ministry of Education, Culture, Sports, Science and Technology of Japan and the Grant-in-Aid of Japan Society for the Promotion of Science (A) 15200037.

References

1. Dowson, D., 1966-67. Modes of lubrication in human joints, Proc. Instn. Mech. Engrs., 181, Pt. 3J, 45-54
2. Murakami, T., 1990. The lubrication in natural synovial joints and joint prostheses, JSME Int. J., Ser. III 33, 465-474.
3. Murakami, T., Higaki, H., Sawae, Y., Ohtsuki, N., Moriyama, S., Nakanishi, Y., 1998. Adaptive multimode lubrication in natural synovial joints and artificial joints, Proc. Instn. Mech. Engrs. 212, Part H, 23-35.
4. Murakami, T., Hayakawa, Y., Higaki, H., Sawae, Y., 1997. Tribological behavior of sliding pairs of articular cartilage and bioceramics, Proc. JSME International Conference on New Frontiers in Biomechanical Engineering, 233-236.
5. Dowson, D., Jin, Z.M., 1986. Micro-elastohydrodynamic lubrication of synovial Joints, Engng. Med., 15, 65-67.
6. Dowson, D., Jin, Z-M., 1987. An analysis of micro-elastohydrodynamic lubrication in synovial joints considering cyclic loading and entraining velocities, In: Dowson, D. et al (Eds.), Fluid Film Lubrication - Osborne Reynolds Centenary. Elsevier, pp.375-386.
7. Swann, D.A., 1978. Macromolecules of synovial fluid, In : Ed. Sokoloff, L. (Ed.), The Joints and Synovial Fluid, vol.I, Academic Press, pp.407-435.
8. Hills, B.A., 1989. Oligolamellar lubrication of joints by surface active phospholipid. J. Rheum. 16(1), 82-91.
9. Higaki, H. and Murakami, T., 1995. Role of constituents in synovial fluid and surface layer of articular cartilage in joint lubrication (Part 2) - The boundary lubricating ability -. Japanese J. Tribology 40 (7), 691-700.
10. Higaki, H., Murakami, T., Nakanishi, Y., Miura, H., Mawatari, T. and Iwamoto, Y., 1998. The Lubricating Ability of Biomembrane Models with Dipalmitoyle Phosphatidylcholine and γ-Globulin. Proc. Instn. Mech. Engrs. 212, Part. H, 337-346.
11. Wong, M, Wuerhrich, P., Buschmann, M.D., Eggli, P. and Hunziker, E., 1997. Chondrocyte Biosynthesis Correlates with Local Tissue Strain in Statically Compressed Adult Articular Cartilage. J. Orthop. Res. 15, 189-196.
12. Guilak, F. and Mow, V.C., 2000. The mechanical environment of the chondrocyte: a biphasic finite element model of cell-matrix interactions in articular cartilage. J. Biomech. 33, 1663-1673.
13. Guilak, F., Anthony, R, and Mow, V.C., 1995. Chondrocyte deformation and local tissue strain in articular cartilage: a confocal microscopy study. J Orthop. Res. 13, 410-421.
14. Schinagl, R.M., Ting, M.K., Price, J.H. and Sah, R.L., 1996. Video microscopy to quantitate the inhomogeneous equilibrium strain within articular cartilage during confined compression. Annals Biomed. Eng. 24, 500-512.

15. Murakami, T., Sawae, Y., Horimoto, M. and Noda, M., 1999. Role of Surface Layers of Natural and Artificial Cartilage in Thin Film Lubrication, In : Dowson, D. et al. (Eds.), Lubrication at Frontier, Elsevier, pp.737-747.

16. Murakami, T., Sawae, Y. and Ihara, M., 2003. The protective mechanism of articular cartilage to severe loading: roles of lubricants, cartilage surface layer, extracellular matrix and chondrocyte. JSME International Journal, Ser.C 46(2), 594-603.

17. Sawae, Y., Murakami, T., Matsumoto, K. and Horimoto, M., 2001. Study on morphology and lubrication of articular cartilage surface with atomic force microscopy. Japanese J.Tribology 45, 51-62.

18. Sasada, T., 2000. Lubrication of human joints. – nature of joint friction and "surface gel hydration lubrication". J. Japanese Soc. for Clinical Biomechanics (in Japanese) 21, 17-22.

19. Ishikawa, I., Murakami, T., Sawae, Y., Miura, H., Kawano, T. and Iwamoto, Y., 2003. Property of Uppermost Superficial Surface Layer of Articular Cartilage as Assessed by Atomic Force Microscopy, Proc. the 4th International Biotribology Forum and the 24th Biotribology Symposium, pp.79-81.

20. Hassan, E.A., Heinz, W.F., Antonik, M.D., D'Costa, N.P., Nageswaran, S., Schoenenberger, C.-A. and Hoh, J.H., 1998. Relative microelastic mapping of living cells by atomic force microscopy. Biophys. J. 74, 1564-1578.

21. Murakami, T., Sakai, N., Sawae, Y. Tanaka, K. and Ihara, M., Influence of proteoglycan on time-dependent mechanical behaviors of articular cartilage under constant total compressive deformation, Submitted to JSME International Journal.

22. Wu, J.Z., Herzog, W. and Epstein, M., 1999. Modelling of location- and time-dependent deformation of chondrocytes during cartilage loading. J. Biomechanics 32, 563-572.

23. Ihara, M., Murakami, T. and Sawae, Y., 2002. Finite element analysis for time-dependent and depth-dependent deformation of articular cartilage and chondrocytes under constant compression. Memoirs of the Faculty of Engineering Kyushu University, 62(4), 165-177.

24. Ihara, M., Murakami, T. and Sawae, Y., 2003, Modelling of articular cartilage containing chondrocytes and finite element analysis for articular cartilage and chondrocytes under compression, Trans JSME (in Japanese), Ser. A 69, No.678, 487-493.

II.   CELL RESPONSE TO MECHANICAL STIMULATION

OSTEOBLASTIC MECHANOSENSITIVITY TO LOCALIZED MECHANICAL STIMULUS DEPENDS ON ORIENTATION OF CYTOSKELETAL ACTIN FIBERS

T. ADACHI

Department of Mechanical Engineering, Graduate School of Engineering,
Kyoto University, Yoshida-Honmachi, Sakyo-ku, Kyoto, 606-8501, Japan
E-mail: adachi@mech.kyoto-u.ac.jp

K. SATO

Department of System Function Science, Graduate School of Science and Technology,
Kobe University, Rokkodai-cho, Nada, Kobe, 657-8501, Japan
E-mail: sato@solid.mech.kobe-u.ac.jp

Osteoblasts play an important role in adaptive bone remodeling under the influence of local mechanical signals such as stress/strain due to loading/deformation. However, the mechanism by which osteoblasts sense the mechanical signals and transduce them into an intracellular biochemical-signaling cascade is still unclear. In this study, calcium-signaling response of the osteoblastic cells, MC3T3-E1, to the local mechanical stimulus was observed by focusing on the involvement of cytoskeletal actin fiber structure in the mechanotransduction pathway. Localized deformation as the mechanical perturbation was applied to a single cell by direct indentation of a glass microneedle, and a change in the intracellular calcium ion concentration $[Ca^{2+}]_i$ was observed as a primal response to the stimulus. The threshold value of the mechanical perturbation was evaluated quantitatively, and its directional dependence due to the aligned cytoskeletal actin fibers was investigated. As a result, mechanosensitivity of the osteoblastic cells to the local mechanical stimulus depends on the angle of the applied deformation with respect to the cytoskeletal actin fiber orientation. This finding is phenomenological evidence that the cytoskeletal actin fiber structures are involved in the mechanotransduction mechanism in osteoblastic cells.

1   Introduction

Bone remodeling is an adaptation process through complex and coordinated series of cellular events including osteoclastic resorption and osteoblastic formation [1]. In this process, physical signals such as stress/strain induced by mechanical loading/deformation play important regulatory roles of cellular activities [2], and lead to intracellular signaling cascades [3, 4]. Although the effects of mechanical stimuli on bone cells have been investigated in *in vitro* experiments by applying fluid shear stress and mechanical strain through their substrate deformation [5-11], little is known about the mechanism by which osteoblasts sense mechanical signals and transmit them into an intracellular signaling response. In those cases, the mechanical stimulus was applied to a population of cultured cells, and average cellular responses were evaluated. Thus, to clarify how a single cell detects and transmits a mechanical stimulus, identification and characterization of the pathway have to be carried out focusing on the cellular mechanical components in a single osteoblastic cell.

Extracellular matrix-integrin-cytoskeletal structure is known as a candidate mechanotransduction pathway [2, 12], and that the cytoskeletal actin fibers influence the cellular response to mechanical stimulus [11]. The cytoskeleton is one of the major intracellular determinants of cellular morphology and functions; integrins are transmembrane receptors that mechanically link cytoskeletal actin-associated proteins to the extracellular matrix. In osteoblastic cells, cytoskeletal actin fibers have a characteristic structure that is oriented along the major axis of the spindle-shaped cell. If this cytoskeletal actin fiber structure plays an important role in the mechanotransduction mechanism in osteoblastic cells, characteristics of the structure may affect the subsequent signaling pathway, which can be observed as a change in a cellular signaling response to a mechanical stimulus.

The aims of this study were to investigate the characteristics of the response of a single osteoblastic cell to a mechanical stimulus due to perturbing the plasma membrane using a glass microneedle, and to examine the involvement of the cytoskeletal actin fiber structure in the mechanotransduction pathway. First, by applying quantitative mechanical perturbation to an osteoblastic cell using a microneedle, a change in the intracellular calcium ion concentration, $[Ca^{2+}]_i$, was observed as the primary signaling response to the mechanical stimulus, through which the threshold value of the perturbation was evaluated quantitatively. Second, to study the directional dependence of the response to the mechanical stimulus, the effect of actin fiber orientation on the threshold value of the calcium signaling response was investigated at various magnitudes and directions of the stimulus.

2 Materials and Methods

2.1 Cell culture and intracellular calcium imaging

Osteoblast-like MC3T3-E1 cells obtained from RIKEN Cell Bank were plated on a glass-bottom dish ($\phi = 35$ mm) at a density of 10^5 cells / dish, cultured in the α-minimum essential medium (α-MEM: ICN Biomedicals) containing 10% fetal bovine serum (FBS: ICN Biomedicals), and maintained in a 95% air and 5% CO_2 humidified environment at 37 deg.C.

Cells were incubated for 3 hours after plating and then loaded with 5 μM fluo-3-AM (Dojindo Molecular Technologies), a fluorescent intracellular Ca^{2+} indicator, in FBS-free α-MEM for 1 hour. Cells were rinsed with phosphate-buffered saline (PBS) and returned again to FBS-free α-MEM.

The change in the intracellular calcium ion concentration, $[Ca^{2+}]_i$, was observed by measuring increases in the fluorescence intensity of fluo-3 using a confocal laser scanning microscope (MRC-1024/MP, Bio-Rad) at room temperature

(23 deg.C). The observed region was 222 x 222 μm^2 under a 60x oil-immersion objective lens. Each fluorescence image, with a size of 512 x 512 pixels, at a height of 2 μm from the dish bottom was scanned at a rate of 1.5 sec per image, and the fluorescence intensity of each pixel was digitized into 8 bits using a computer.

2.2   Cell orientation

Fluorescence and transmitted images of an osteoblastic cell are superimposed and shown in Fig. 1A, in which the tip of the microneedle can be seen at the center of the cell. In osteoblastic cells, cytoskeletal actin fibers have an aligned structure whose direction significantly coincides with the major axis of the spindle-shaped cell ($p < 0.01$, $r = 0.83$ for $n = 20$), in which the cellular axis was determined as the major axis of an ellipse fitted with the cell shape determined using an image processing software (Image-Pro Plus, Media Cybernetics), and the major axis of cytoskeletal actin fibers was determined as that of the fabric ellipse [14] measured for a Rhodamine-Phalloidin fluorescent image of the actin fibers. The angle between the microneedle and the cell axis was defined as θ (deg.), as illustrated in Fig. 1B.

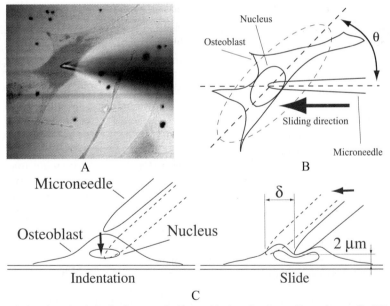

Figure 1. Local mechanical stimulus to a single osteoblastic cell using a glass microneedle [13]. A: Superimposed fluorescence and transmitted images of a cell and the tip of a microneedle. B: Definition of angle θ between the cell axis and the direction of applied deformation. C: Schematic of deformation applied to a single cell using the tip of a microneedle and definition of applied perturbation δ.

2.3 Mechanical stimulation

A mechanical stimulus was applied to a single osteoblastic cell using a tip of a glass microneedle with a tip diameter of 10 μm. The tip of microneedle was heated to make it smooth and round. The microneedle was attached to a three-dimensional hydraulic micromanipulator (MHW-103, Narishige) at an angle of 40 deg. between the microneedle axis and the dish plane.

The schematic of the deformation applied to a single cell is shown in Fig. 1C. First, the tip of the microneedle was moved down vertically to indent the cell surface to a height of 2 μm from the dish bottom, and held there for a few seconds, as illustrated in Fig. 1C (left); note the thickness of the cell at the point of indentation was about 6 to 8 μm. Second, after confirming the lack of cellular response to the indentation, the microneedle was moved horizontally in the direction of angle θ (Fig. 1B) at a speed of 10 μm/s to deform the cell, as illustrated in Fig. 1C (right), where the displacement of the tip was defined as δ (μm).

3 Results

3.1 Response of osteoblastic cell to mechanical stimulus

A transient increase in the intracellular calcium ion concentration, $[Ca^{2+}]_i$, in a single osteoblastic cell in response to the applied mechanical stimulus was observed as an increase in the fluorescence intensity of fluo-3, as shown in Fig. 2A. Figure 2B shows the time course change in the average fluorescence intensity in the cell, in which arrows (a) to (d) correspond to the cells in Figs. 2A(a) to (d), respectively. Figure 2A(a) shows the fluorescent image of the cell before stimulation. When the tip of microneedle was indented vertically at the center of the cell, a shadow of the tip was observed as a black spot (white arrow) on a focus plane at the height of 2 μm from the dish bottom, as shown in Fig. 2A(b). After confirming that this indentation did not induce significant change in $[Ca^{2+}]_i$, the tip of the microneedle was displaced horizontally, $\delta = 8$ μm, to deform the cell at $t = 0$ s, as shown in Fig. 2A(c). Immediately after the stimulation, the fluorescent intensity increased and spread within the cell, subsequently reached its peak value at $t = 12$ s (Fig. 2A(d)), and then decreased gradually toward the basal level before stimulation, as shown in Fig. 2B. Once the calcium response was observed, the microneedle was moved away from the cell.

When the local deformation was applied to a single osteoblastic cell using a microneedle, some cells responded with increase in $[Ca^{2+}]_i$ and some did not. In order to examine the effect of the magnitude of deformation on cellular response, the displacement magnitude δ in the horizontal direction was varied from 2 μm to 12 μm at 2 μm intervals. The percentage of cells P that responded to the stimulus is analyzed against the applied deformation δ for cells ($n = 39$) in the range from $\theta =$

60 to 90 deg (data not shown) as an example. For small displacements, $\delta = 2$ to 6 μm, no cell responded to the mechanical stimulation. However, with increase in the applied displacement, the percentage of cells that responded to the stimulus P increased, and 80 % of the cells responded to the large displacement $\delta = 12$ mm. This result indicates that there is a threshold of displacement that cells can sense and respond to the mechanical stimulus.

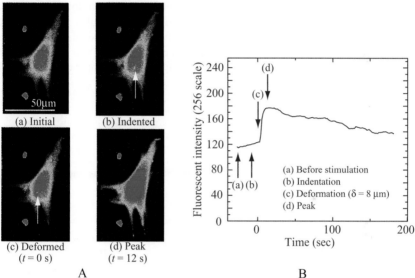

Figure 2. Calcium signaling response to perturbation applied to a single osteoblastic cell using a glass microneedle [13]. A: Fluorescent images of responded cells. B: Change in fluorescent intensity in time.

3.2   Directional dependence of response to mechanical stimulus

From observations in the previous section, it was indicated that a cell has a threshold value at which it can respond to a mechanical stimulus. If the cytoskeletal actin fiber structure plays an important role in the mechanotransduction mechanism in osteoblastic cells, characteristics of the structure may affect the threshold value at which it responds to a mechanical stimulus. Here, we focus on the orientation of the cytoskeletal actin fiber structure in the osteoblastic cells that may cause a directional dependence of the response to a mechanical stimulus. To examine the directional dependence of the response on the threshold value, the above-mentioned cellular calcium response to the mechanical stimulus was studied by applying deformation in various directions θ.

Cells were divided into three groups based on the angle θ, Group A (◆), $\theta = 0 \sim 30$ deg ($n = 35$); Group B (■), $\theta = 30 \sim 60$ deg ($n = 35$); and Group C (▲), $\theta = 60 \sim 90$ deg ($n = 39$). The percentage of the cells P that responded is plotted in Fig.

3A against the magnitude of applied displacement δ. As can be seen in this figure, for smaller displacements $\delta = 2 \sim 6$ μm, no cells responded to the stimulation in all groups. For displacement $\delta = 8$ μm, cells in Group A did not respond, while 25 % and 44 % of the cells in Group B and Group C responded to the stimulation. For displacement $\delta = 1$ μm, 14 %, 50 %, and 63 % of the cells in Groups A, B, and C responded, respectively. In each group, it can be observed that a larger displacement leads to a higher percentage of response. In addition, for displacement $\delta = 8$ μm and 10 μm, a larger angle θ seems to lead to a higher percentage of response. These results indicate that the sensitivity of an osteoblastic cell to a mechanical stimulus is affected by both the magnitude and direction of displacement, that is, the sensitivity seems to be related to the angle of cytoskeletal actin fiber orientation.

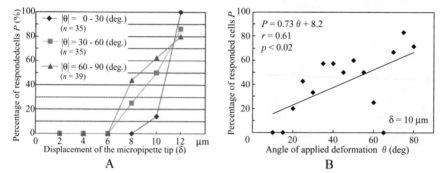

Figure 3. Percentage of responded cells to mechanical stimulus [13]. A: Three groups divided based on the angle of applied deformation. B. Directional dependence of cellular response for $\delta = 10$ μm.

To investigate directional dependence of the osteoblastic response to a mechanical stimulus, experimental data were analyzed by plotting the percentage of cells P that responded against the applied deformation angle θ. Figures 3B present the dependence of the angle of applied deformation θ on the percentage of the responded cells P for the applied displacements $\delta = 10$ μm, where the percentage was plotted as a function of the angle of applied deformation by taking moving average within 10 deg at 5 deg intervals.

For displacement $\delta = 8$ μm (data not shown), the percentage of the cellular response P increased with the increase in the applied deformation angle θ. Linear regression analysis revealed a correlation coefficient $r = 0.68$ with P and θ significantly correlated ($p < 0.01$). For $\delta = 10$ μm, linear regression analysis showed a significant positive correlation ($p < 0.02$) between P and θ with a coefficient $r = 0.61$, as shown in Fig. 3B. These results suggest a directional dependence of the threshold value at which osteoblastic cells respond to the mechanical stimulus, which implies that the sensitivity to the mechanical stimulus depends on the direction relative to the cytoskeletal actin fiber orientation.

4   Discussion

To clarify the osteoblastic response to a mechanical stimulus, that is, to answer questions such as what kinds of stimuli activate cellular activities and how are these signals related to intracellular structural components such as the cytoskeletal actin fiber structure, it is desirable to setup experimental conditions isolated from complex factors other than mechanical stimulation as much as possible. Osteoblasts communicate with neighboring cells via an intercellular network through direct contacts called gap junctions [15] and indirect signaling mechanisms. For example, propagation of intercellular calcium signaling is one possible mechanism through which a cell can affect other cells to coordinate their activities [16]. Thus, when a mechanical input is applied *in vitro* to cells contacting each other, we cannot isolate the effect of the mechanical stimulus itself on the cellular response from that of these intercellular communications. In this study, to exclude these complex factors arising from neighboring cells, the mechanical stimulus was quantitatively applied to a single osteoblastic cell using a microneedle.

No calcium signaling response was observed for a small deformation by a microneedle, but a larger deformation caused a higher percentage of the response. This displacement magnitude dependence indicates the existence of a threshold value in sensing a mechanical stimulus in osteoblastic cells, which reminds us of a tissue remodeling response with a threshold value of strain around the remodeling equilibrium [17]. However, the applied displacement δ in this *in vitro* experiment cannot be directly transformed into an overall strain such as the strain applied to the cell through the substrate deformation, because the deformation due to the microneedle displacement was very localized in the vicinity of the microneedle tip; that is, this experiment induces a complex and heterogeneous strain field in the cell thus it is not possible at present to correlate the response to strain magnitude. Thus, for quantitative comparison with the *in vivo* experimental data, controlled local stress or strain measure should be applied to the cell as a mechanical input.

In osteoblastic cells, the cytoskeletal actin fiber structure tends to align along the major axis of cell shape. This cytoskeletal structure is postulated to play an important role in the mechanotransduction pathway through which extracellular mechanical signals such as force and deformation are transduced into the intracellular signaling cascade [18]. From these two viewpoints, the experiment was conducted based on the hypothesis that the characteristics of the mechano-transduction mechanism are affected by the orientation of the cytoskeletal actin fiber structure. As a result, it was found that changing the angle between the applied deformation using a microneedle and the cellular orientation resulted in a directional dependence of the intracellular calcium response to the mechanical stimulus.

One possible model that interprets the directional dependence is that the difference in the apparent membrane stiffness due to the aligned fibers beneath the membrane may affect the localized deformation of the plasma membrane. This anisotropic stiffness due to fiber alignment may cause a variation in the local membrane stretch depending on the direction, and the directional dependence could be observed if the stretch-activated channel was involved in this mechanical response. Another model is the case that the cytoskeletal actin fibers are directly playing a role as the component of mechanical signal transduction. If the critical tensile deformation along the fiber direction exist that follows to the next signaling cascade through the fiber structure itself, the tensile deformation along the fiber may be the key component that is geometrically determined from the applied deformation and its direction with respect to the fiber axis.

The stretch-activated channel [19, 20] is a candidate involved in this experiment to cause the change in $[Ca^{2+}]_i$. For example, deformation applied to the plasma membrane of cultured osteoblasts using a glass microneedle induced an increase in $[Ca^{2+}]_i$ that was significantly attenuated by the ion-channel blocker on the plasma membrane [16]. Thus, the increase in $[Ca^{2+}]_i$ observed in this study due to the microneedle perturbation might be also considered as a result of the ion flux into the cell through the channel on the plasma membrane. To clarify in more detail the role of the cytoskeletal actin fiber structure in the mechano-transduction pathway and the relation to the ion channels, experimental conditions have to be set up such that the activation of stretch-activated channel and of other possible factors is prevented.

Cells change their morphology to align along specific directions under uniaxial mechanical stimulations such as fluid shear stress and substrate stretching. This is well documented for endothelial cells under flow [21] and stretching [22, 23]. Similar to this, osteoblastic cells were reported to change their orientation under stretching [24]. In these cells, it is considered that the cell shape and structure are altered or maintained, as mediated by the cytoskeletal actin fiber structure that links to the extracellular matrix via surface focal adhesion receptors such as integrins [23]. These mechanical components are involved in the cell remodeling process wherein they change their distribution and configuration in the cell [25]. However, how cells sense anisotropic mechanical stimuli such as force and displacement vectors, and how mechanical inputs induce the polarized distribution of the components are still unknown. To adapt their shape dynamically to directional mechanical inputs, cells must have some mechanism though which they can sense directional mechanical inputs and distribute their components to construct oriented cytoskeletal structures and polarized focal adhesions.

The observed directional dependence of the osteoblastic response to microneedle perturbations is phenomenological evidence that the cytoskeletal actin fiber structure is involved in mechanotransduction mechanism, which might be related to cellular polarized behavior. Even though the present results are still a phenomenological observation of the involvement of the cytoskeletal actin fiber structure in the mechanotransduction process in osteoblastic cells, the approach from a mechanical viewpoint is another step toward obtaining insights into the mechanotransduction mechanism and to construct a basis of studies in the field of bone mechanobiology.

Acknowledgements

This work was partially supported by Grant-in-Aid for Scientific Research on Priority Areas 15086211 from the Ministry of Education, Culture, Sports, Science and Technology of Japan.

References

1. Parfitt, A. M., 1994, Osteonal and hemi-osteonal remodeling: The spatial and temporal framework for signal traffic in adult human bone, J. Cell. Biochem., 55-3, 273-286.
2. Duncan, R. L. and Turner, C. H., 1995, Mechanotransduction and the functional response of bone to mechanical strain, Calcif. Tis. Int., 57, 344-358.
3. Banes, A. J., Tsuzaki, M., Yamamoto, J., Fischer, T., Brigman, B., Brown, T., and Miller, L., 1995, Mechanoreception at the cellular level: The detection, interpretation, and diversity of responses to mechanical signal, Biochem. Cell Biol., 73-7/8, 349-365.
4. Turner C. H. and Pavalko, F. M., 1998, Mechanotransduction and functional response of the skeleton to physical stress: The mechanisms and mechanics of bone adaptation, J. Orthop. Sci., 3, 346-355.
5. Yeh, C.-K., Rodan, G. A., 1984, Tensile forces enhance prostaglandin E synthe-sis in osteoblastic cells grown on collagen ribbons, Calcif. Tis. Int., 36, S67-71.
6. Hasegawa, S., Sato, S., Saito, S., Suzuki, Y., and Brunette, D. M., 1985, Mechanical stretching increases the number of cultured bone cells synthesizing DNA and alters their pattern of protein synthesis, Calcif. Tis. Int., 37, 431-436.
7. Buckley M. J., Banes A. J., Levin L. G., Sumpio B. E., Sato M., Jordan R., Gilbert J., Link G. W., and Tran Son Tay, R., 1988, Osteoblasts increase their rate of division and align in response to cyclic, mechanical tension *in vitro*, Bone & Min., 4, 225-236.
8. Neidlinger-Wilke, C., Wilke, H. J., and Claes, L., 1994, Cyclic stretching of human osteoblasts affects proliferation and metabolism: A new experimental method and its application, J. Orthop. Res., 12, 70-78.
9. Jacob, C. R., Yellowley, C. E., Davis, B. R., Zhou, Z., Cimbala, J. M. and Donahue, H. J., 1998, Differential effect of steady versus oscillating flow on bone cells, J. Biomech., 31, 969-976.
10. Kurata, K., Uemura, T., Nemoto, A., Tateishi, T., Murakami, T., Higaki, H., Miura, H., and Iwamoto, Y., 2001, Mechanical strain effect on bone-resorbing activity and messenger RNA expressions of marker enzymes in isolated osteoclast culture, J. Bone & Min. Res., 16-4, 722-730.

11. Adachi, T., Murai, T., Hoshiai, S., and Tomita, T., 2001, Effect of actin filament on deformation-induced Ca^{2+} response in osteoblast-like cells, JSME Int. J., 44C-4, 914-919.

12. Ingber, D., 1991, Integrins as mechanochemical transducers, Cur. Opin. Cell Biol., 3, 841-848.

13. Adachi, T., Sato, K., and Tomita, Y., 2003, Directional dependence of osteoblastic calcium response to mechanical stimuli, Biomech. Model. Mechanobiol., 2, 73-82.

14. Cowin, S. C., 1985, The relationship between the elasticity tensor and the fabric tensor, Mech. Mat., 4, 137-147.

15. Donahue, H. J., McLeod, K. J., Rubin, C. T., Andersen, J., Grine, E. A., Hertzberg, E. L., and Brink, P. R., 1995, Cell-to-cell communication in osteoblastic networks: Cell line-dependent hormonal regulation of gap junction function, J. Bone & Min. Res., 10-6, 881-889.

16. Xia, S.-L. and Ferrier, J., 1992, Propagation of a calcium pulse between osteoblastic cells, Biochem. Biophys. Res. Comm., 186-3, 1212-1219.

17. Frost, H. M., 1987, The mechanostat: A proposed pathogenic mechanism of osteoporoses and the bone mass effects of mechanical and nonmechanical agents, Bone & Min., 2-2, 73-85.

18. Meazzini, M. C., Toma, C. D., Schaffer, J. L., Gray, M. L., and Gerstenfeld, L. C., 1998, Osteoblast cytoskeletal modulation in response to mechanical strain in vitro, J. Orthop. Res., 16-2, 170-180.

19. Guharay, F. and Sachs, F., 1984, Stretch-activated single ion channel currents in tissue-cultured embryonic chick skeletal muscle, J. Physiol., 352, 685-701.

20. Sokabe, M., Sachs, F., and Jing, Z.-Q., 1991, Quantitative video microscopy of patch clamped membranes stress, strain, capacitance, and stretch channel activation, Biophys. J., 59-3, 722-728.

21. Levesque, M. J. and Nerem, R. M., 1985, The elongation and orientation of cultured endothelial cells in response to shear stress, Trans. ASME, J. Biomech. Eng., 107-4, 341-347.

22. Naruse, K., Yamada, T., and Sokabe, M., 1998, Involvement of SA channels in orienting response of cultured endothelial cells to cyclic stretch, Am. J. Physiol., 274-5, H1532-1538.

23. Wang, J. H.-C., Goldschmidt-Clermont, P., Wille, J., and Yin, F. C., 2001, Specificity of endothelial cell reorientation in response to cyclic mechanical stretching, J. Biomech., 34-12, 1563-1572.

24. Wang, J. H.-C., Grood, E. S., Florer, J., and Wenstrup, R., 2000, Alignment and proliferation of MC3T3-E1 osteoblasts in microgrooved silicone substata subjected to cyclic stretching, J. Biomech., 33, 729-735.

25. Sokabe, M., Naruse, K., Sai, S., Yamada, T., Kawakami, K., Inoue, M., Murase, K., and Miyazu, M., 1997, Mechanotransduction and intracellular signaling mechanisms of stretch-induced remodeling in endothelial cells, Heart Vessels, 12S, 191-193.

MICROBIOMECHANICAL PROPERTIES OF CULTURED ENDOTHELIAL CELLS ESTIMATED BY ATOMIC FORCE MICROSCOPY

M. SATO

Department of Biomedical Engineering and Robotics, Tohoku University,
6-6-01 Aoba-yama, Sendai 980-8579, Japan
E-mail: sato@bml.mech.tohoku.ac.jp

Local mechanical properties were measured for bovine endothelial cells exposed to shear stress using an atomic force microscopy (AFM), and the AFM indentations were simulated by a finite element method (FEM) to determine the elastic modulus. After exposure to shear stress, the endothelial cells showed marked elongation and orientation in the flow direction, together with significant decrease in the peak cell height. The applied force-indentation depth curve was obtained at different locations of the cell surface and quantitatively expressed by the quadratic equation. The elastic modulus was determined by comparison of the experimental and numerical results. The modulus obtained in our FEM model significantly became higher from 12.2 ± 4.2 kPa to 18.7 ± 5.7 kPa with exposure to shear stress. Fluorescent images showed that stress fibers of F-actin bundles were mainly formed in the central portion of the sheared cells. The significant increase in the modulus may be due to the remodeling of cytoskeletal structure. The elastic modulus would contribute a better understanding of the mechanisms of endothelial cell remodeling processes during exposure to shear stress.

1 Introduction

Vascular endothelial cells change their macroscopic shape, microstructure and physiological functions in response to fluid shear stress [1, 2]. We have observed that intracellular F-actin filament distributions changed depending on the shear stress and the flow direction [3, 4] and have suggested that stress distributions in the cells might be also accompanied by the reorganization of cytoskeletal structures [5]. In order to discuss the intracellular stress distributions, we need to determine local mechanical properties in the cells.

Indentation tests using the atomic force microscope (AFM) have potential to measure detailed microbiomechanical properties of soft biological samples, including platelets, endothelial cells, and epithelial cells. In these studies, local indentations were performed on the cells to estimate elastic modulus, but the results are still controversial: which is stiffer on the nucleus or the peripheral regions of the cells? This difference may be due to experimental parameters, such as the indentation depth, the indentation velocity, the fitting curve in the force-indentation curve and so on. Some researchers [6-8] have developed finite element model to investigate the effects of the geometry of the cantilever tip, the indentation depth, specimen size and specimen nonlinearity and inhomogeneity. A newly designed AFM has been developed to measure local mechanical properties in combination with observation of microstructure for endothelial cells [4, 9]. It will be shown in this paper that the stiffness of the cells exposed to shear stress increases with the duration time of exposure. Furthermore, we apply the finite element method (FEM) to cell deformation by the AFM indentation and determine the local elastic modulus of sheared endothelial cells [10]. The FEM model is also used to survey the effects of the relative specimen geometry on the force-indentation curve.

2 Methods

2.1 Endothelial cell culture

Bovine aortic endothelial cells (BAECs) were obtained from thoracic aortas and cultured with Dulbecco's modified Eagle medium (Gibco Laboratories, MD, USA) supplemented with 10% heat-inactivated fetal bovine serum (JRH Biosciences, KS, USA), penicillin, and streptomycin (Gibco) at 37°C 95% air/5% CO_2, as described previously [3-5]. After reaching confluence after incubation for 4 to 5 days, BAECs were trypsinized and then plated in a cell culture dish with a diameter of 35 mm (Asahi Techno Glass, Chiba, Japan). BAECs with passages 4 to 9 were used for flow-imposed experiments.

2.2 Fluid-imposed experiments

Fluid flow was applied to confluent monolayers of BAECs using a parallel-plate flow apparatus similar to that described in detail elsewhere [3-5]. Briefly, the parallel plate flow unit was loaded into the cell culture dish, in which the flow channel has 0.5 x 14 mm rectangular section. A steady shear stress of 2 Pa for 6 - 24 h was applied to the BAEC monolayers by perfusing the same culture medium at 37°C in 95% air/5% CO_2 through the channel using a roller pump. Statically cultured cells were maintained in an incubator as controls.

2.3 AFM measurements

Local mechanical properties of BAECs were measured with a custom-built AFM apparatus as shown in Fig. 1 [9]. A unique feature of this apparatus is that this system can allow us to visualize cytoskeletal structure simultaneously together with AFM measurements because the AFM is combined with an inverted confocal LSM (LSM-GB200, Olympus, Tokyo, Japan).

After exposure to shear stress, the three-dimensional topography of BAECs was first obtained by the AFM and then some locations on the major axis of the cell surface were selected for the measurement of the local mechanical response being the cantilever tip placed over. The cantilever has a length of 200 μm and a spring constant of 0.02 N/m (Olympus, Tokyo, Japan). The force-indentation curve was obtained for each location with the indentation depth of 0-500 nm and the indentation rate of 880 nm/s. The applied force F and the resulting indentation depth δ curve was expressed by the quadratic equation:

$$F = a\delta^2 + b\delta \tag{1}$$

where a and b are the parameters expressing the nonlinearity and the initial stiffness of the force-indentation curve, respectively. The elastic modulus was obtained by comparison of the experimental and numerical results. The numerical analysis will be shown below.

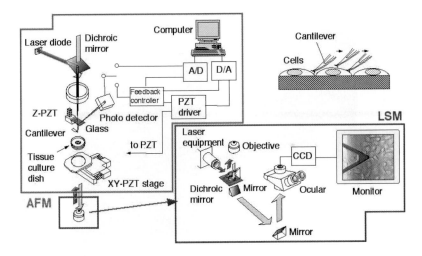

Figure 1. Schematic drawing of the optical system of the newly designed AFM combined with the confocal laser scanning microscope. X-, Y-, Z-PZT are piezo actuators for x, y, z directions, respectively. [9]

In a separate study [11], the LSM was used to observe F-actin filament distributions of BAECs. BAECs were fixed with 10% formaldehyde for 5 min and then stained with rhodamine-phalloidin at a dilution in sterile phosphate buffered saline of 1:20 for 20 min. Fluorescence images were observed through a water immersion x60 (1.0 NA) objective at room temperature and transferred to a personal computer.

2.4 Finite element method

Prior to the AFM experiments, the FEM analysis was performed. For ease of implementation, we constructed an axisymmetric FEM model to simulate cantilever indentation by AFM, of which a typical mesh is shown in Fig. 2. All nodes on the lower surfaces were constrained both axially and radially. The model consisted of a rigid element representing the cantilever, a gap element representing the contact region between the cantilever and the specimen surface, and 2185 of an isoparametric quadrilateral element used to descretize the specimen with the smallest elements located under the indenter. A conical indenter with tip semi-angle $\alpha = 45$ deg was used. The cantilever was assumed to be rigid comparing with the specimen and the specimen was assumed to be homogeneous, isotropic, and linear elastic. Poisson's ratio of 0.49 was used to approximate incompressibility. Indentation was imposed by applying a uniformly distributed pressure to the cantilever surface vertically. The resulting force F and indentation δ curve is thus obtained to compare with its counterpart which is independently obtained through the AFM experiments.

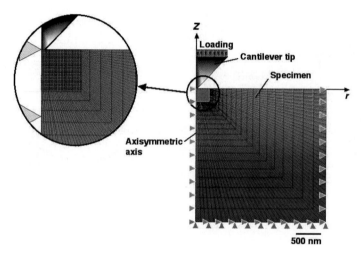

Figure 2. Axisymmetric finite element mesh. The model consists of the AFM cantilever and the specimen. Reprinted from [10] with permission from IOS Press.

The matching process to estimate the elastic modulus E_{FEM} using our FEM model is schematically shown in Fig. 3. Corresponding linear equation can be derived from the analysis in the following form:

$$\frac{F}{E_{FEM}} = b' \cdot \delta \tag{2}$$

where b' is the linear coefficient. The force-indentation curve obtained from geometrically nonlinear analysis is also shown in the figure. The elastic modulus E_{FEM} was determined so as to identify the linear coefficient b' of Eq. (2) and its experimental counterpart, that is, the initial slope b of Eq. (1) obtained from the experiments. Finally we obtained the following relationship.

$$E_{FEM} = \frac{b}{b'} \qquad (3)$$

The analysis was also performed to survey the effects of the relative specimen geometry on the force-indentation curve by varying the specimen thickness and radius. The present finite element analysis was conducted by using the ANSYS 5.5 commercial code (SAS IP, PA, USA) running on a workstation (Ultra 10, Sun Microsystems, CA, USA).

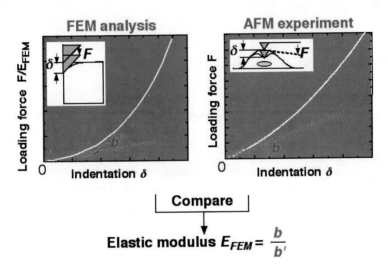

Figure 3. Estimation process of the elastic modulus. The elastic modulus E_{FEM} is determined so as to identify the linear coefficient b' in the FEM analysis and the coefficient b in the AFM experiment. Reprinted from [10] with permission from IOS Press.

2.5 Data analysis and statistics

Data are shown as mean ± SD; n represents the number of experiments performed on different cells. Statistical comparisons were made by use of unpaired Student's t test. A difference was considered to be significant at a value of $P < 0.05$.

3 Results

Figure 4 shows typical photomicrographs of rhodamine-phalloidin stained endothelial cells [11]. After exposure to shear stress, the cells show marked elongation and orientation with the flow direction. Thick stress fibers are mainly formed in the central portion of the cells and aligned with the flow direction.

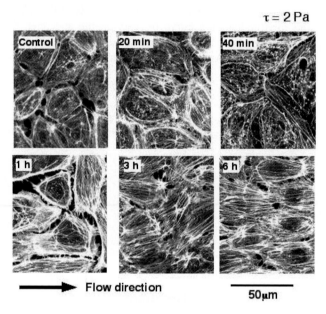

Figure 4. Typical photomicrographs of rhodamine-phalloidin stained endothelial cells before and after exposure to shear stress. [11]

The indentation test was also performed on endothelial cells exposed to a shear stress for period of 6 and 24 h. The regions of upstream and downstream sides were found to be stiffer than the centre region. The values for both parameters, a and b, are summarized in Fig. 5 for the total number of endothelial cells exposed to shear stress for 6 and 24 h. The control (C) data are also included for comparison. It is evident that sheared endothelial cells become much stiffer than the corresponding group of statically cultured cells. At 6-h exposure, parameter, a, becomes larger, which is equivalent to an increase in stiffness at the upstream side. At 24-h exposure, there is no difference in mechanical properties between upstream, center and downstream locations of the cell.

Typical contour map for sheared endothelial cell measured by AFM and the locations (1-7) where the indentation measurement was carried out are shown in Fig. 6. The three-dimensional geometries of the cells were constructed by scanning the cell surface at a constant interaction force of less than 0.1 nN. The peak cell height

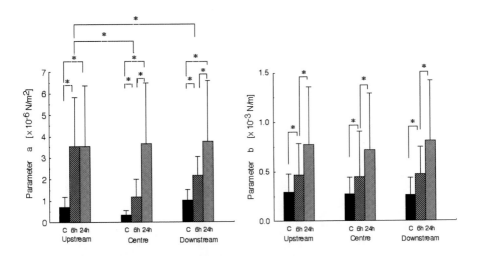

Figure 5. Parameters, a and b, obtained at upstream, center and downstream sides in statically cultured control (C) and in sheared endothelial cells for 6 and 24 h. Control: n=48, 6h: n=21, 24h: n=22. Mean±SD. * p<0.05. Reprinted from [4] with permission from Elsevier.

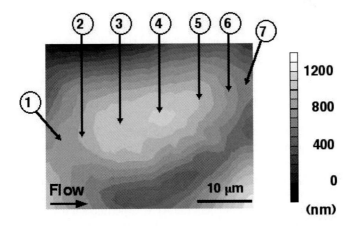

Figure 6. Typical contour map measured by AFM for a sheared endothelial cell (2 Pa, 24 h) and measured locations. The numbers of 1 to 7 are the locations where the indentation test was carried out.

significantly decreases 2.8 ± 1.0 μm to 1.4 ± 0.5 μm with fluid flow. The elastic modulus E_{FEM} calculated for control and sheared cells using our model is shown in Fig. 7. The elastic moduli E_{FEM} are 12.2 ± 4.2 kPa (mean±SD) for control and 18.7 ± 5.7 kPa (mean±SD) ($P < 0.05$ vs. control) for sheared cells, which are directly derived from the parameter *b*. The modulus significantly increased with fluid shear stress.

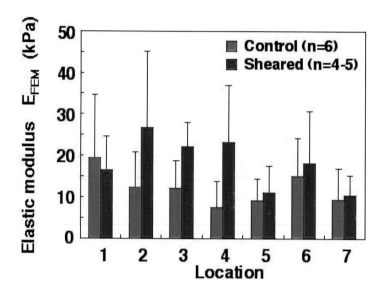

Figure 7. Elastic modulus E_{FEM} using our FEM model for control and sheared cells. The modulus significantly increases from 12.2±4.2 kPa for control to 18.7 kPa ($p<0.05$) for sheared cells. Reprinted from [10] with permission from IOS Press.

4 Discussion

In this study FEM analysis was performed to analyze AFM indentations and the local mechanical properties of bovine endothelial cells exposed to shear stress were estimated.

The FEM analysis indicated that the effects of specimen size can be neglected if the specimen thickness and width is appropriate size [10]. Karduna *et al.* [7] performed FEM analysis to supplement the AFM experimental data and showed the useful guidelines for the indentation tests such as boundary effects, showing good agreement with our results. This result implies that one should use the force-indentation curve up to an extent region according to the predicted cell thickness. Since the cell thickness at the peripheral regions is very small, the measured modulus may be affected by the specimen boundary. Costa and Yin [8] also used FEM models to examine the effects of indentation depth and material nonlinearity together with indenter geometry on the finite indentation response. They have shown that the two critical factors determining an apparent elastic modulus are whether the deformations are infinitesimal and whether the material exhibits nonlinear characteristics. In this analysis, the parameter b, *i.e.* initial stiffness, is used to estimate the elastic modulus, which has little effect on both the geometrical and material nonlinearities. The tip geometry may arise some errors: the actual geometry of the AFM tip is not conical but pyramidal. Despite these difficulties, the present analysis deserves further consideration.

In Fig. 5, both the parameters a and b increased for sheared endothelial cells, which indicates the remodeling in cytoskeletal structure. Fluorescent images showed that thick stress fibers of F-actin bundles were observed for sheared endothelial cells, as shown in Fig. 4. The increase in the parameters, a and b, may be due to this cytoskeletal remodeling. Figure 7 shows a significant increase in the elastic modulus E_{FEM} with exposure to shear stress. Theret *et al.* [12] applied a pipette aspiration technique to bovine endothelial cells exposed to shear stress of 1 and 3 Pa for 4 to 24 h to study mechanical properties, and showed an increase in the elastic modulus with fluid shear stress. However our results are ten times or more higher than their values. One reason for this discrepancy is due to a difference in measurements between AFM and pipette aspiration. Another reason is that we used the attached cells on the dish while their cells were the suspended ones. From this point of view, there should be significant difference in cytoskeletal structures between the flatten cells and the rounded cells.

In summary, we have demonstrated that the FEM was combined with the AFM indentations to determine the local mechanical properties of sheared endothelial cells. The elastic modulus using our FEM model significantly increased with exposure to shear stress, showing higher values in our model. Combination of FEM and AFM measurements allows us to measure accurate local mechanical properties and might contribute to new insights into the stiffness of the cytoskeleton.

Acknowledgments

I wish to thank Drs. T. Matsumoto (present: Nagoya Institute Technology), N. Kataoka (present: Kawasaki Medical College) and T. Ohashi for their superb researches adopted in this paper. This work was supported financially in part by Grants-in-Aids for Scientific Research 14208100 and 15086203 from the Ministry of Education, Culture, Sports, Science and Technology, Japan and the Asahi Glass Foundation.

References

1. Levesque, M.J., Nerem, R.M., 1985. The elongation and orientation of cultured endothelial cells in response to shear stress. Trans. ASME J. Biomech. Eng. 107, 341-347.

2. Nerem, R.M., Levesque, M.J., Cornhill, J.F., 1981. Vascular endothelial morphology as an indicator of the pattern of blood flow. Trans. ASME J. Biomech. Eng. 103, 172-176.
3. Kataoka, N., Ujita, S., Sato, M., 1998. Effect of flow direction on the morphological responses of cultured bovine aortic endothelial cells. Med. Biol. Eng. Comput. 36, 122-128.
4. Sato, M., Nagayama, K., Kataoka, N., Sasaki, M., Hane, K., 2000. Local mechanical properties measured by atomic force microscopy for cultured endothelial cells exposed to shear stress. J. Biomech. 33, 127-135.
5. Ohashi, T., Sugawara, H., Matsumoto, T., Sato, M., 2000. Surface topography measurement and intracellular stress analysis of cultured endothelial cells exposed to fluid shear stress. JSME Intl. J. Ser. C 43, 780-786.
6. Zhang, M., Zheng Y.P., Mak, A.F.T., 1997. Estimating the effective Young's modulus of soft tissues from indentation tests-Nonlinear finite element analysis of effects of friction and large deformation. Med. Eng. Phys. 19, 512-517.
7. Karduna, A.R., Halperin, H.R., Yin, F.C.P., 1997. Experimental and numerical analyses of indentation in finite-sized isotropic and anisotropic rubber-like materials. Annals Biomed. Eng. 25, 1009-1016.
8. Costa, K.D., Yin, F.C.P., 1999. Analysis of indentation: implication for measuring mechanical properties with atomic force microscopy. Trans. ASME J. Biomech. Eng. 121, 462-471.
9. Nagayama, K., Sasaki, M., Hane, K., Matsumoto, T., Sato, M., 2004. Development of newly designed atomic force microscope system for measuring of mechanical properties of cells. Trans. Jap. Soc. Mech. Eng. Ser. C 70(691), 736-742.
10. Ohashi, T., Ishii, Y., Ishikawa, Y., Matsumoto, T., Sato, M., 2002. Experimental and numerical analyses of local mechanical properties measured by atomic force microscopy for sheared endothelial cells. Bio-Med. Mat. and Eng. 12, 319-327.
11. Kataoka, N., Sato, M., 1998. The change of F-actin distribution and morphology of cultured bovine aortic endothelial cells in the early stage of fluid shear stress exposure. Trans. Jap. Soc. Mech. Eng. Ser. B 64(622), 1801-1808.
12. Theret, D.P., Levesque, M.J., Sato, M., Nerem, R.M., Wheeler, L.T., 1988. The application of a homogeneous half-space model in the analysis of endothelial cell micropipette measurements. Trans. ASME J. Biomech. Eng. 110, 190-199.

EFFECTS OF MECHANICAL STRESSES ON THE MIGRATING BEHAVIOR OF ENDOTHELIAL CELLS

[1]T. TANAKA, [1,2]K. NARUSE AND [1,2,3]M. SOKABE*

[1]*Department of Physiology, Nagoya University Graduate School of Medicine, [2]ICORP Cell Mechanosensing, JST, 65 Tsurumai, Nagoya 466-8550, Japan, [3]Deaprtment of Molecular Physiology, National Institute for Physiological Sciences, NINS, Okazaki 444-8585, Japan*
Tatsuya Tanaka: twin@po2.synapse.ne.jp
Keiji Naruse: knaruse@med.nagoya-u.ac.jp
*Masahiro Sokabe (corresponding author): msokabe@med.nagoya-u.ac.jp

Objective: Endothelial cells are exposed to a variety of mechanical stresses, which modulate a number of endothelial functions. One of the important functions of endothelial cells is their migrating ability displayed at healing of vascular injury and angiogenesis. The present study aimed to elucidate the effects of mechanical stresses on the migrating behavior of bovine aorta endothelial cells (BAECs). **Methods:** BAECs were cultured on a fibronectin-coated elastic silicone membrane. A narrow scar (*ca.* 200 μ m in width) was made by a scratch in the confluent monolayer of BAECs, and migrating behavior of remaining cells into the vacant area was measured under controlled mechanical stresses onto the cells. **Results:** When the silicone membrane was continuously stretched by 20% at a stroke perpendicular to the scar, cell migration was strongly accelerated. In contrast, when a 20% prestretched silicone membrane in the same axis as above was relaxed, which effectively generated compressive force onto the cells, cell migration toward the vacant area was significantly inhibited. When mechanical stresses were applied parallel to the scar, cell migration was accelerated moderately by either stretch or compression. Direction of migration and lamellipodia formation was also affected significantly by mechanical stresses. **Conclusion:** Migrating behavior of BAECs is influenced not only by the modes (stretch or compression) of the applied mechanical stress but also by its orientation (perpendicular or parallel to the scar).

1   Introduction

Endothelial cells are exposed to a variety of mechanical stresses including blood pressure, shear stress, and circumferential tension originated from pulsatile blood flow. These mechanical stresses induce a variety of responses in endothelial cells. Shear stress activates cytoskeletal remodeling, reorganization of extracellular matrix, and synthesis of specific proteins [1, 2]. Cyclic stretch exerts similar effects [1, 3]. Expression level of certain genes is also modulated by mechanical stresses [4-7]. However, the signaling mechanism of these mechanically induced cell responses was remained largely unknown mainly because of the complexity of mechanotransduction processes in the cell. Currently, stretch activated (SA) channel, phospholipase C, adenylate cyclase, Na^+/H^+ exchanger, volume sensitive channel, G protein, tyrosine kinases, cytoskeleton, and integrin are supposed to be involved in the cellular mechanotransduction process [1, 7-12].

Migration of endothelial cells plays a critical role in angiogenesis and healing of vascular injury. A number of studies have been performed on the effects of shear stress in the migration of endothelial cells, where the migration was found to be enhanced by proper shear stress [2, 10, 13]. However, few studies have been conducted on the effects of stretch or compression, another important stress in cells, mainly because of the technical difficulty in applying controlled mechanical forces onto the cell.

In this study, we developed a method by which we can apply mechanical forces onto the cell with different modes (stretch and compression) and orientations in a controlled manner. BAECs were cultured on an elastic thin silicone membrane and a narrow scar (ca 200 µm in width) was made by a scratch in their confluent monolayer. The migrating behavior of BAECs into the denuded area was analyzed under various mechanical stresses created by stretch (or its release) of the membrane. It was found that these mechanical forces affected significantly on the extent and the direction of BAECs migration. Noticeably, when the stresses were applied perpendicularly to the scar, directed migration of the cells toward the vacant area was enhanced *ca.*5 times as the control by stretch, while it was inhibited nearly completely by compression.

2   Materials and Methods

2.1   Cell culture

BAECs were prepared according to the method by Ryan et al [14]. Cells were cultured in a 5% CO_2 incubator at 37℃ with 80% Eagle's minimum essential medium (Nissui Pharmaceutical Co. Japan) supplemented with $NaHCO_3$, L-gultamine, 20% fetal calf serum, 50 µIC/ml penicillin, and 50 µg/ml streptomycin. Endothelial cells were identified by morphological features. Subcultured cells with the passage number of 3 to 10 were used for the experiments.

2.2   Application of mechanical stresses to the cells

Elastic silicone membranes (Fuji system Japan, 200 µm in thickness) were coated for 30 minutes with 50 µg/ml fibronectin (Takaken Japan) diluted with culture medium. Subcultured BAECs were plated on the membrane at the density of 10^6 /ml. As schematized in Figure 1, two opposite edges of a silicone membrane were clamped mechanically with custom-made mechanical holders. To apply quantitative mechanical strain to the silicone membrane, one holder was fastened tightly and the other was connected to a micromanipulater. After cells were cultured confluently, medium was changed to a standard extracellular solution (SES: 145mM NaCl, 5mM KCl, 2mM $CaCl_2$, 2mM $MgCl_2$, 10mM Gulcose, 10mM HEPES, pH 7.40 adjusted with NaOH). All chemicals used were of special grade (Wako Japan). Then a narrow scar (*ca.* 200 μm in width) was made by a scratch with a plastic pippet tip (Gilson 1-200 μl), and spontaneous migration of remaining cells into the vacant area was measured in SES. This method is known as the wound assay [15]. A time series of typical cell migration toward vacant area is shown in Figure 2. After the scar was made, cells were washed 3 times with SES to exclude the effects of substances from injured cells [16]. There was no significant difference in migration activity between on fibronectin coated membranes and on scratched membranes. Then cells were subjected to continuous stretch at a stroke parallel (Figure 1, left) or perpendicular (Figure 1, right) to the scar by stretching the silicone membrane from 20mm to 24mm (defined as 20% stretch). To apply compressive forces to the cells, the silicone membrane was stretched from 20mm to 24mm before cells were cultured on it, and relaxed to the original length just before the migration measurement (defined as 20% compression).

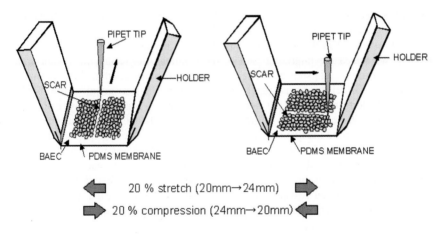

Figure 1. Schematic diagram of experimental setup. BAECs were cultured confluently on an elastic silicone (PDMS) membrane. Two opposite edges of the silicone membrane were clamped mechanically with metal holders. One holder was fastened tightly and another was connected to a micromanuplater to apply uni-axial mechanical strain to the silicone membrane. After making a narrow scar (*ca.* 200 μm in width) by a scratch with a pipet tip, mechanical strain was applied and cell migration into the vacant area was measured. To apply tensile forces to the cells, the silicone membrane was stretched from 20mm to 24mm, while compressive forces were applied by relaxing prestretched membrane from 24mm to 20mm. The orientation of mechanical stresses was controlled by changing the orientation of the scar as shown above.

2.3   Analysis of cell migration

Cells under controlled mechanical stresses were mounted on the stage of a phase contrast inverted microscope (Nikon Diaphoto). The microscope was covered with an incubator housing (Nikon NP-2), and temperature inside was kept at 37°C.

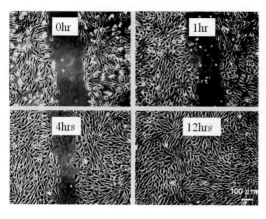

Figure 2. A time series of phase contrast images showing cell migration toward denuded area (a vertically running dark region). These images were taken from the cells cultured on a glass cover slip but not on a silicone membrane to give a better cell images. Therefore this is not an actual result we analyzed in this study, but rather for a demonstration to show how the wound-healing model works.

The cells moved towards the vacant area with the formation of lemellipodia. Migrating cells were continuously observed on a TV monitor with the analysis field of 195 μm x 270 μm through a 20x objective and a video camera (Ikegami ICD-42DC). We traced the migrating path of the frontier cells facing the vacant area. Generally frontier cells showed elongated shape with lamellipodia. We determined long axis of the cell by visual inspection and traced the movement of the middle point of the long axis every 30 minutes for 6 hours (Figure 3). Most cells kept contact with other cells during migration, but some cells detached from other cells and began to migrate independently. We excluded such isolated cells from the data. The migration path of the cells without mechanical stresses was not straight toward the vacant area but rather showed a zig-zag way. However, under mechanical stresses, cells tended to move in a particular direction. To quantify this behavior of the cells, we employed two parameters as the indices of migration, the migration range and the effective distance (Figure 3). The former value, which represents total

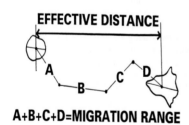

EFFECTIVE DISTANCE

A+B+C+D=MIGRATION RANGE

Figure 3. Schematic diagram showing how we analyzed the cell migration. Cells were mounted on the stage of a microscope and phase contrast images of migrating cells were continuously recorded on a video-tape. Usually migrating cells showed elongated shape with lamellipodia. We determined long axis of the cell and traced the movement of the middle point of the long axis every 30 minutes for 6 hours. Two parameters were employed to estimate the migrating activity of the cells; the migration range and the effective distance. The former value was used to describe the migration activity of the cells, and the latter the effective migration activity to cover the vacant area. The ratio between the effective distance and the migration range was defined as the efficiency to recover the vacant area.

distance of the migration path, reflects the total activity of cell migration. The latter value, which represents the shortest distance from the original frontier line, reflects the effective migration activity to cover the vacant area. The ratio between the effective distance and the migration range was defined as the efficiency of migration to recover the vacant area. Higher efficiency means shorter time to cover the vacant area. We adopted the migration range, the effective distance, and the efficiency at 1 hour in the estimation of mechanical effects on the cell migration, because effects of mechanical stresses became maximum at 1 hour as mentioned in results. To estimate the allover effects of mechanical stresses during the course of obervation (6 hours), we employed the effective distance because physiologically relevant aspect of the cell migration in this study is the recovery of vacant area. In addition effects of mechanical stresses were found to be more apparent in the effective distance than in the migration range. The effective distance every 30 minutes was used as a parameter of migration velocity. The present investigation conforms to the guide for the care and use of laboratory animals published by the US National Institutes of Health.

3 Results

3.1 Cell migration toward the vacant area without external mechanical forces

Soon after scratching the confluent cell layer, frontier cells facing the vacant area began to move with forming lamellipodia. Cells behind the frontier cells also began to change their shape and moved gradually toward the vacant area. Recovery of the vacant area was established not only by the migration of frontier cells but also by the movements of all the remaining cells behind frontier cells. This observation suggested that individual cells migrated depending on the gradient of intercellular mechanical stresses. The margin of frontier cells facing the vacant area is free of external stress, and the stress at the margin of the cells just behind frontier cells will be decreased as frontier cells move to the vacant area. The margin of lamellipodia was moving back and forth and finally extension of lamellipodia in a specific direction was established. Cells showed elongated shape in the direction of lamellipodia formation and some of such cells detached from other cells. These isolated cells began to migrate faster than the cells attached to others. Without external mechanical forces, direction of lamellipodia formation and migration was relatively random, and the velocity of migration was almost constant during the course of observation for 6 hours.

3.2 Effects of mechanical stresses applied perpendicularly to the scar

When 20% stretch was applied perpendicularly to the scar, cell migration into the vacant area was strongly accelerated. In contrast cell migration was inhibited seriously by 20% compression in the same axis (Figure 4). Stretching the membrane must have increased the tension in the cell surface attached on the membrane, and increased the traction force at the leading edge of migrating cells. In contrast releasing the membrane stretch (compression) would decrease the tension in the cell and dècrease the traction force at the leading edge facing the vacant area. As shown in Figure 5, velocity of migration in response to stretch was higher in the first 2 hours and decreased as time passed. This result suggests that effects of stretch applied *via* substratum became weaker as cells moved on it. To evaluate the effects of mechanical stresses on the direction of migration, we took the efficiency (effective distance / migration range: see methods) at 1 hour, because the velocity of migration, which may reflect the actual effects of mechanical stresses, became maximum at 1 hour as shown in Figure 5.

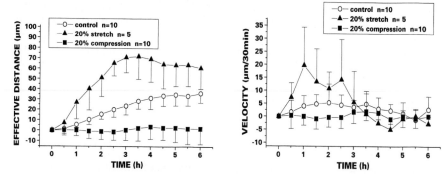

Figure 4 (left panel). Effects of mechanical stresses applied perpendicularly to the scar on the effective distance. Effective distances were measured every 30 min under constant external stress applied perpendicularly to the scar. Mean values ± SD are shown in the figure for control (n=10 O), 20% stretch (n=5 ▲), and 20% compression (n=10 ■).

Figure 5 (right panel). Effects of mechanical stresses applied perpendicularly to the scar on the velocity of migration. Increments of effective distance per 30 min (defined as the velocity of migration here) were measured under constant external stress. This is actually a differential expression of Figure 5. Mean values ± SD are shown in the figure for contorol (n=10 O), 20% stretch (n=5 ▲), and 20% compression (n=10 ■).

As summarized in Table 1, the migration range (in μm) at 1 hour, which represents migration activity of the cell, was significantly enhanced by 20% stretch (33.50), whereas slightly changed by 20% compression (13.65) compared with control (13.43). The effective distance (in μm) at 1 hour was much larger at 20% stretch (27.03), and significantly lower at 20% compression (0.29) compared with control (5.58). As the result, the efficiency at 1 hour was enhanced significantly by 20% stretch (0.76), whereas attenuated severely by 20% compression (0.03) compared with control (0.44). These results clearly indicate that migrating behavior of endothelial cells is strongly influenced by the mode of externally applied mechanical stresses. Fig. 6 illustrates representative migrating behavior of the picked out three cells in different mechanical conditions. Under the control condition (without mechanical stresses), the direction of lamelipodia formation was random. At 20% stretch perpendicular to the scar, the lamelipodia formation toward the scar was accelerated appreciably. At 20% compression, in contrast, the lamelipodia formation toward the scar was suppressed remarkably.

3.3   Effects of mechanical stresses applied parallel to the scar

Above observation suggests that the orientation of applied forces critically influences the direction of cell migration in the wound-healing model. To test this possibility, we examined the effects of mechanical stresses parallel to the scar on the cell migration into the scar. Either 20% stretch or 20% compression applied parallel to the scar accelerated cell migration (Fig. 7). This makes a good contrast with the stretch and compression applied perpendicular to the scar, which caused opposite effects on cell migration (Fig. 4).

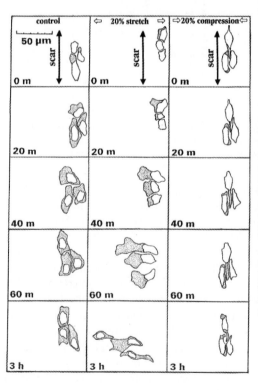

Figure 6.   Migration and lamellipodia formation of frontier cells facing the scar (indicated by a double headed arrow in the top raw) under various mechanical stresses: left column, control (no stress); middle, 20% stretch perpendicular to the scar; right, 20% compression perpendicular to the scar. In each column appearances of cell shape and lamellipodia formation are shown at 0min, 20min, 40min, 60min, and 3hr after scratch. Lamellipodia formation was apparent in the first 1 hour. Lamellipodia, which are shown as dotted areas in individual cells, could be distinguished from cell soma region (open areas) in phase contrast images. Without mechanical stresses, direction of lamellipodia formation was relatively random. In contrast, under 20% stretch, lamellipodia formation and migration in the direction of stretch are accelerated remarkably while they are largely suppressed under 20% compression.

Table 1. Migration indices under various mechanical conditions. Migration indices (effective distance, migration range, and efficiency.) here were taken at 1 hour after the onset of cell migration since the migration velocity became maximum at 1 hour as shown in Figures 5 and 8. Values are indicated as mean values ± SD.

| | Control | Perpendicularly | | Parallel | |
| | | Stretch | Compression | Stretch | Compression |
	N=10	N=5	N=10	N=5	N=10
Effective Distance (μm)	5.58±2.86	27.03±16.11	0.29±5.40	18.03±9.60	12.96±2.93
Migration Range (μm)	13.43±5.24	33.50±15.67	13.65±5.31	19.70±8.07	18.55±3.73
Efficiency	0.44±0.16	0.76±0.32	0.03±0.41	0.87±0.16	0.71±0.16

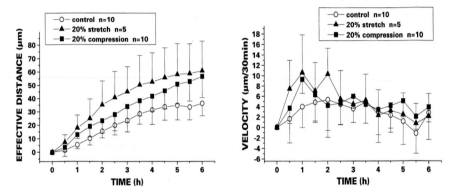

Figure 7 (left panel). Effects of mechanical stresses applied parallel to the scar on the effective distance. Effective distances were measured every 30 min under constant stress applied parallel to the scar. Mean values ± SD are shown in the figure for control (n=10 ○), 20% stretch (n=5 ▲), and 20% compression (n=10 ■).

Figure 8 (right panel). Effects of mechanical stresses applied parallel to the scar on the velocity of migration. Increments of effective distance per 30 min (velocity of migration) were measured under constant stress. This is actually a differential expression of Figure 7. Mean values ± SD are shown in the figure for control (n=10 ○), 20% stretch (n=5 ▲), and 20% compression (n=10 ■).

This result clearly shows that cell migration in response to mechanical stresses depends on the orientation of mechanical stresses. The velocity of migration, which was enhanced by the stresses parallel to the scar, was highest in the first 1 to 2 hours and then decreased with time (Figure 8). As summarized in table 1, the migration range (in µm) at 1 hour was enhanced by either 20% stretch (19.70) or 20% compression (18.55) compared with control (13.43). Similarly the effective distance and efficiency at 1 hour were enhanced either by stretch or compression parallel to the scar, suggesting that mechanical stresses other than the compression perpendicular to the scar (parallel to the direction of cell migration) accelerate cell migration.

4 Discussion

4.1 Cell migration without external forces

The present study showed that endothelial cells migrated toward the denuded area and that the migration was significantly influenced by externally applied mechanical forces depending on their modes (stretch or compression) and orientations (perpendicular or parallel to the scar). Before discussing underlying mechanisms of these mechanical effects, we consider here the cell migration without external mechanical forces. The cell migration toward the vacant area is triggered most likely by a release of intrinsic mechanical stress (pressure) from adjacent cells at the edge facing the vacant area. Spatial gradient of the pressure in the axis perpendicular to the scar may underlie the directed cell movement. Supportive observations were reported on the effect of shear stress, where cell migration was activated more in downstream than in upstream [2, 13]. In the confluent cells, pressure among the cells may inhibit cell migration.

4.2 Cell migration under mechanical stresses

An application of mechanical stretch in a direction perpendicular to the scar resulted in a significant acceleration of cell migration toward the denuded area. At the leading edge of migrating cells, lamellipodia formation was greatly accelerated as shown in Figure 6. Although precise mechanisms are not fully understood, polymerization of actin filaments at the cell margin, which will cause an internal pressure to push the margin, is supposed to play a pivotal role in the lamellipordia formation [17]. The mechanical stretch could have augmented the polymerization of actin filaments. The inhibition of cell migration induced by compression perpendicular to the scar can be interpreted by a reverse effect on the actin polymerization. Alternatively, the compressive force applied might have acted as a direct force to prevent the cell movement. The cell migration toward the vacant area was enhanced either by stretch or compression when the stress was applied in a direction parallel to the scar. In the case of compression, the pressure around the cell margin except for the leading edge facing the vacant area may have increased, giving rise to a force that would accelerate the cell migration toward the scar. It is more difficult to explain the effects of stretch. It could be due to an alteration of intracellular Ca^{2+} dynamics as addressed below.

4.3 Involvement of stretch activated ion channels in cell migration

The dramatic effects of mechanical stresses on the cell migration presented here indicate that cells can sense not only the amplitude but also the direction of mechanical stresses. Considering that continual deformation of the cell during migration will cause changes in the mechanical stresses in the cell, cell mechanosensing mechanisms would play an essential role in the control of migration even without external mechanical forces. The only identified cell mechanosensing machinery to date is the stretch activated (SA) ion channel that is assumed to be activated by the membrane tension generated by cell membrane deformation [18]. SA channels have been identified in many cell types and reported to play a key role in mechanically induced cell responses [7, 18-20]. A Ca^{2+} permeable SA channel was found in porcine aorta endothelial cells [21]. We also found a similar SA channel in human umbilical vein endothelial cells, which could increase the intracellular Ca^{2+} concentration in response to stretch [22]. It is highly possible that similar SA channels are expressed in our preparation(BAECs) and involved in cell migration.

To test this hypothesis we made a series of preliminary experiments under the conditions without external mechanical forces. We used gadolinium(Gd^{3+}), which was first discovered to block an SA channel in Xenopus oocytes [23] and has been used as a potent blocker for SA channels. Also Gd^{3+} is known to inhibit various cell responses to mechanical stresses [3, 24] and the stretch induced intracellular Ca^{2+} increase *via* SA channels in endothelial cells [22]. Application of Gd^{3+} (50-100 μM) caused a significant and dose-dependent inhibition of the cell migration toward the denuded area (not shown). The Gd^{3+} effects were similar to those of a depletion of Ca^{2+} from the external solution. These observations suggest that Ca^{2+} influx through SA channels play an important role in the BAEC migration.

4.4 Sensing the direction of mechanical stress and clinical implications

The second messenger, intracellular Ca^{2+}, is known to induce a variety of migration-related responses. Ca^{2+} dependent protease is needed in detaching cells from substratum [25]. Lamellipodia formation underlying migration is regulated by a variety of actin binding proteins, some of which are Ca^{2+} dependent [26, 27]. A precise model of cell migration suggests a localized increase in intracellular Ca^{2+} concentration during migration [28], and a spatial gradient of Ca^{2+} concentration was reported in migrating eosinophils [29]. Such a spatial gradient of intracellular Ca^{2+} concentration may be caused by local Ca^{2+} influx through SA channels. We can speculate that localized Ca^{2+} influx through SA channels regulates dynamics of cytoskeletons through Ca^{2+} dependent actin regulating proteins, leading to a directed cell migration. Another potential candidate for the mechanosensing machinery is the adhesion contact [1, 10-12], which is a complex structure consisting of integrin and various associated proteins, some of which are liked to cytoskeletal structures mostly stress fibers. Mechanical stresses in the substratum will be conveyed through adhesion molecules to cytoskeletons, and may directly affect the actin polimerization and directed lamellipodial development. The possible involvement of cytoskeletons in mechanosensing would be of great interest. As the cytoskeleton is a filamentous structure with a particular direction, it may work as an antenna to detect the direction of mechanical stresses in the cell. If the molecular complex constituted of cytoskeleton, integrin and SA channels works as a cell mechanosensor, it may be able to detect the direction of mechanical forces. There are reports indicating the importance of cytoskeletal networks in the activation of SA channels [9, 18, 30].

Our results showed that the migration of BAECs was greatly influenced by mechanical stresses depending on their modes (stretch or compression) and orientations (parallel or perpendicularly to the scar). In other words, endothelial cells are equipped with such a fine mechanosensing machinery that can discriminate not only stretch and compression but also their orientations. It was surprising that stretch and compression exerted completely opposite effect on the cell migration when they were applied perpendicularly to the scar. However, the phenomenon looks very reasonable since an open wound at stretched region must be healed more rapidly. Such a mode- and orientation-dependent cell migration would also enhance efficiency in the angiogenesis as well as in the healing process of endothelial injury.

Acknowledgements

The authors thank Prof. Kodama (Nagoya University) for his critical reading of our manuscript. This work was supported by Grants-in-Aid for Scientific Research 13480216, Scientific Research on Priority Areas 15086270, and Creative Scientific Research 16GS0308 from the Ministry of Education, Culture, Sports, Science and Technology of Japan, and a grant from Japan Space Forum to MS.

References

1. Lehoux, S., Tedgui, A., 1998. Signal transduction of mechanical stresses in the vascular wall. Hypertension 32, 338-345.
2. Tardy, Y., Resnick, N., Nagel, T., 1997. Gimbrone MA Jr, Dewey CF Jr. Shear stress gradients remodel endothelial monolayers in vitro via a cell proliferation-migration-loss cycle. Arterioscler Thromb Vasc. Biol. 17,3102-3106.
3. Naruse, K., Yamada, T., Sokabe, M., 1998. Involvement of SA channels in orienting response of cultured endothelial cells to cyclic stretch. Am. J. Physiol. 274, H1532-H1538.
4. Chien, S., Li, S., Shyy, Y.J., 1998. Effects of mechanical forces on signal transduction and gene expression in endothelial cells. Hypertension 31, 162-169.
5. Kanzaki, M., Nagasawa, M., Kojima, I., Sato, C., Naruse, K., Sokabe, M., Iida, H., 1999. Molecular identification of a eukaryotic, stretch-activated nonselective cation channel. Science 285, 882-886.

6. Naruse, K., Sai, X., Yokoyama, N., Sokabe, M., 1998. Uni-axial cyclic stretch induces c-src activation and translocation in human endothelial calls via SA channel activation. FEBS Lett. 441, 111-115.

7. Sokabe, M., Naruse, K., Sai, S., Yamada, T., Kawakami, K., Inoue, M., Murse, K., Miyazu, M., 1997. Mechanotransduction and intracellular signaling mechanisms of stretch-induced remodeling in endothelial cells. Heart Vessels, 12, 191-193.

8. Baraka, A.I., 1999. Responsiveness of vascular endothelium to shear stress: potential role of ion channels and cellular cytoskeleton. Int. J. Mol. Med. 4, 323-332.

9. Chen, C.S., Ingber, D.E., 1999. Tensegrity and mechanoregulation: from skeleton to cytoskeleton. Osteoarthritis Cartilage 7, 81-94.

10. Rainger, G.E., Buckley, C.D., Simmons, D.L., Nash, G.B., 1999. Neutrophils sense flow generated stress and direct their migration through $\alpha_v \beta_3$ integrin. Am. J. Physiol. 276, H858-864.

11. Traub, O., Berk, B.C., 1998. Laminar shear stress: mechanisms by which endothelial cells transduce an atheroprotective force. Arterioscler Thromb Vasc. Biol. 18, 677-685.

12. Wang., N., 1998. Mechanical interactions among cytoskeletal filaments. Hypertension 32, 162-165.

13. Ando, J., Nomura, H., Kamiya, A., 1987. The effect of fluid shear stress on the migration and proliferation of cultured endothelial cells. Microvasc Res. 33, 62-70.

14. Ryan, U.S., Mortana, M., Whitaker, C., 1980. Methods for microcarrier culture of bovine pulmonary artery endothelial cells avoiding the use of enzymes. Tissue Cell 12, 619-635.

15. Burk, R.R., 1973. A factor from a transformed cell line that affects cell migration. Proc. Natl. Acad. Sci. USA 70, 369-373.

16. McNeil, P.L., Muthukrishnan, L., Warder, E., D'Amore, P.A., 1989. Growth factors are released by mechanically wounded endothelial cells. J. Cell. Biol. 109, 811-822.

17. Abraham, V.C., Krishnamarthi, V., Taylor, D.L., Lanni, F., 1999. The actin-based nanomachine at the leading edge of migrating cells. Biophys. J. 77, 1721-1732.

18. Sokabe, M., Sachs, F., Jing, Z., 1991. Quantitative video microscopy of patch clamped mambranes:stress,strain,capacitance,and stretch channel activation. Biophys. J. 59, 722-728.

19. Guharay, F., Sachs, F., 1984. Stretch-activated single ion channel currents in tissue-cultured embryonic chick skeletal muscle. J. Physiol. 352, 685-701.

20. Sokabe, M., Sachs, F., 1992. Towards molecular mechanism of activation in mechanosensitive ion channels. Adv. Compa. Envi. Physiol. 10, 55-77.

21. Lansman, J.B., Hallam, T.J., Rink, T.J., 1987. Single stretch-activated ion channels in vascular endothelial cells as mechanotransducers? Nature 325, 811-813.
22. Naruse, K., Sokabe, M., 1993. Involvement of stretch-activated ion channels in Ca^{2+} mobilization to mechanical stretch in endothelial cells. Am. J. Physiol. 264, C1037-C1044.
23. Yang, X-C, Sachs, F., 1989. Block of stretch-activated ion channels in xenopus oocytes by gadolinium and calcium ions. Science 243, 1068-1071.
24. Swerup, C., Purali, N., Rydqvist, B., 1991. Block of receptor response in the stretch receptor neuron of the crayfish by gadolinium. Acta. Physiol. Scand. 143, 21-26.
25. Leavesley, D.I., Schwartz, M.A., Rosenfeld, M., Cheresh, D.A., 1993. Integrin β_1- and β_3- mediated endothelial cell migration is triggered through distinct signaling mechanisms. J. Cell Biol. 121, 163-170.
26. Glogauer, M., Arora, P., Yao, G., Sokholov, I., Ferrier, J., McCulloch, C.A.G., 1997. Calcium ions and tyrosine phosphorylation interact coordinately with actin to regulate cytoprotective responses to stretching. J. Cell Sci. 10, 11-21.
27. Stossel, T.P., 1989. From signal to pseudopod. J Biol Chemi 264, 18261-18264.
28. Lee, J., Ishihara, A., Theriot, J.A., Jacobson, K., 1993. Principles of locomotion for simple-shaped cells. Nature 362,167-171.
29. Brundage, R.A., Fogarty, K.E., Tuft, R.A., Fay, F.S., 1991. Calcium gradients underlying polarization and chemotaxis of eosinophils. Science 254, 703-706.
30. Janmey, P.A., 1998. The cytoskeleton and cell signaling: component localization and mechanical coupling. Physiol. Rev. 78, 763-781.

III. TISSUE ENGINEERING

ENGINEERING APPROACHES TO REGULATE CELL DIFFERENTIATION AND TISSUE REGENERATION

T. USHIDA AND K. S. FURUKAWA

Biomedical Engineering Laboratory, Graduate School of Engineering, University of Tokyo,
7-3-1 Hongo, Tokyo 113-8656, Japan
E-mail: ushida@mech.t.u-tokyo.ac.jp

G. CHEN AND T. TATEISHI

Biomaterials Center, National Institute for Materials Science,
1-1 Namiki,Tsukuba 305-0044, Japan
E-mail: Guoping.CHEN@nims.go.jp

Methodology for cell differentiation and tissue regeneration in vitro is keenly required towards the adaptation to regenerative medicine. The methodology could be composed from three indispensable approaches other than biochemical one. One is an approach from material side, where cells could receive outside-in signals from biomaterials such as collagen, hyaluronan. 3D scaffolds hybridizing biomaterials with biodegradable polymers could be served as an incubator not only for controlling cell differentiation but also for promoting tissue regeneration. The second is physical approach, where cells also could transduce outer physical stimulations into intracellular signals. Hydrostatic pressure, shear stress could be used for regulating differentiation of chondrocytes, endothelial cells, and for regenerating articular cartilage, blood vessel, respectively. The last is one from cell-cell interaction, which is known to be necessary in the developmental steps. Spheroids, as far as necrosis is avoided, are useful for controlling cell differentiation, and could become micro-tissue elements for tissue regeneration.

1   Introduction

It is one of the most crucial themes in regenerative medicine to regulate cell differentiation and tissue regeneration in vitro. Both cell differentiation and tissue regeneration are thought to be limited under in vitro conditions. Especially, we could not discuss tissue regeneration, apart from tissue necrosis. In vitro culture requires novel methodologies to realize tissue nutrition by any means such as angiogenesis.

Knowledge about cell differentiation and tissue regeneration accumulates by contribution from biochemical fields. It is reasonable to adopt such knowledge in cell biology and molecular biology into regenerative medicine. As a ligand interacts its specific receptor and evokes cellular signal transduction, biochemical stimulations such as growth factor addition are estimated to be an adequate method to regulate cell differentiation and tissue regeneration.

On the other hand, it is known for cells to be regulated by intracellular signals evoked by other than biochemical stimulation. Cell adhesion assemblies involving

92

integrins are well known to regulate cell adhesion, spreading and migration by inside-out signals, and regulate at the same time cell differentiation and tissue regeneration by outside-in signals, which are caused by integrin-matrix interaction. Another interaction regulating cell differentiation and tissue regeneration is cell-cell interaction, which is realized for example in cell aggregation at developmental steps. Thus, an approach is presumed probable to regulate cell differentiation and tissue regeneration by means of realizing cell-matrix interaction and cell-cell interaction from the engineering point of view.

Moreover, it is known that physical stimulation in addition to biochemical one evokes intracellular signal transduction and diverse cellular responses. Blood vessel is physiologically loaded with shear stress by blood flow, and tensile stress by pulsatile blood pressure. Femur bone is loaded with compressive stress and articular cartilage is loaded with hydrostatic pressure by walking and weighting. Those physical stimulations are known to activate diverse cascades of intracellular signals at the same time. Although their effects are thought not to be specific in comparison with biochemical one, the adoption of those physical stimulations to regulate cell differentiation and tissue regeneration could be one of the effective engineering approaches.

As described-above, three approaches could be thought to realize cell differentiation and tissue regeneration in vitro from the engineering point of view; 1) approach using three dimensional scaffolds, 2) approach using physical stimulation, 3) approach using cell-cell interaction, as shown in Fig.1.

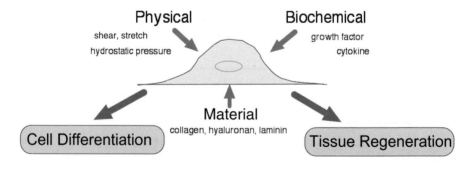

Figure 1.   Three essential approaches for regulating cell differentiation and tissue regeneration.

2   Approach to Regulate Cell Differentiation and Tissue Regeneration by Means of Three Dimensional Scaffolds

Cartilage is one of the promising candidates for tissue regeneration in vitro by using tissue engineering, because chondrocytes could endure low oxygen concentration and nutrition supply. It is necessary for regenerated cartilage to be bigger than one taken with biopsy in order to exceed mosaic plasty. For that purpose, it is inevitable to make chondrocytes proliferated in vitro. Articular cartilage is composed of type II collagen, although fibrous cartilage is composed of type I collagen. However, it is well recognized that chondrocytes are dedifferentiated according to proliferation in vitro, producing not type II but type I collagen. Therefore, tissue engineering is requested to establish a new methodology how to make dedifferentiated chondrocytes redifferentiated in vitro.

Biomaterials derived from tissues such as collagen, hyaluronan have been used in tissue engineering. Those materials have sites to be able to interact with integrins or receptors on the plasma membrane and evoke intracellular signals, although they are relatively weak in mechanical properties. On the other hand, biodegradable polymers such as poly glycolic acid (PGA), poly lactic acid (PLLA), and their copolymer poly DL-lactic-co-glycolic acid (PLGA) have also been frequently used in tissue engineering. Those polymers have enough mechanical strength, although they are relatively hydrophobic and have no sites to interact with cells.

In order to compensate mutually demerits of biomaterials and biodegradable polymers, their hybridization has been tried. For that context, we have made the hybridization of collagen micro-sponges with PLGA sponges [1-3], the hybridization of collagen micro-sponges with PLGA meshes [4], the hybridization of collagen micro-sponges with hydroxyapatite micro-beads and PLGA sponges [5], and the hybridization of collagen gels including chondrocytes with PLLA non-woven fiber scaffolds [6]. Thus, those hybridization could promote not only efficiencies of cell seeding and cell adhesion but also regeneration ability of cartilage-like tissues in vitro. Fig. 2 shows SEM micrograph of PLGA-collagen hybrid mesh (a), phase contrast micrograph of two passaged cells immediately after seeding in the hybrid mesh (b), and SEM micrographs of chondrocytes cultured in the hybrid mesh for 1 week (c) and 4 weeks (d). The hybrid mesh was prepared by forming cobweb-like collagen micro-sponges in the openings of a knitted mesh made of PLGA (Fig. 2a). Observation by a phase contrast microscope indicates that the cells were entrapped by the collagen micro-sponges in the hybrid mesh immediately after cell seeding (Fig. 2b). Chondrocytes adhered to the cobweb-like hybrid mesh and showed uniform distribution on the mesh. They proliferated and regenerated cartilaginous matrices, filling the void spaces in the hybrid mesh (Fig. 2c,d). Fig.2 also shows Northern blot analysis of the genes encoding type I collagen, type II collagen, and aggrecan of two passaged chondrocytes cultured in the hybrid mesh for 0, 2, 4, and 12 weeks. When the dedifferentiated chondrocytes were seeded and cultured in the hybrid mesh, the expression of mRNAs for type II collagen and aggrecan was up-regulated and that of type I collagen mRNA was down-regulated. After culture in the hybrid mesh for 12 weeks, the gene expression of type I collagen was very weakly detectable, but those of type II collagen and aggrecan reached their highest levels. Those results show a possibility for such hybridized scaffolds to possess the function enabling dedifferentiated chondrocytes

94

redifferentiated in coincidence with cartilage-like tissue regeneration [7]. It could be also thought possible to regulate cell differentiation and tissue regeneration in more complicated tissue such as blood vessel and bone, using signals from matrices and three dimensional structures of those hybridized scaffolds [8, 9].

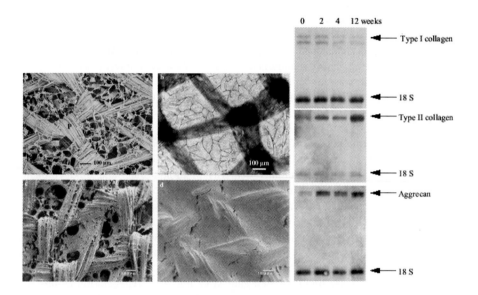

Figure 2.   SEM micrograph of PLGA-collagen hybrid mesh (a), phase contrast micrograph of two passaged cells immediately after seeding in the hybrid mesh (b), and SEM micrographs of chondrocytes cultured in the hybrid mesh for 1 week (c) and 4 weeks (d). (left)   Northern blot analysis of the genes encoding type I collagen, type II collagen, and aggrecan of two passaged chondrocytes cultured in the hybrid mesh for 0, 2, 4, and 12 weeks (right) (Ref.7).

3   Approach to Regulate Cell Differentiation and Tissue Regeneration by Means of Physical Stimulation

Cells and tissues are known to be physiologically loaded with diverse physical stimulations. For example, femur is loaded with compressive or tensile stress by walking, which is known to evoke micro-deformation, streaming potential in bone tissue. In blood vessel, endothelial cells are loaded with shear stress by blood flow, endothelial cells and smooth muscle cells are loaded with tensile stress by blood pulse. Those physical stimulations are known to evoke intracellular signals such as Ca, cAMP. Therefore, trials have been done to regenerate three-dimensional tissues by mimicking physical stimulations.

Articular cartilage is loaded with compressive stress by walking and exercise. Cartilage is a tissue with high water-content, due to charged polymers such as proteoglycans, other than a hard tissue with high stiffness. Therefore, chondrocytes are thought to be loaded with hydrostatic pressure caused by compressive stress. According to in vivo measurements, hip articular cartilage is loaded with 3-7 MPa of hydrostatic pressure by walking. The hydrostatic pressure has been tried to be used to regulate cellular functions of chondrocytes. If hydrostatic pressure is generated by compressing medium through air-phase, the concentrations of gases such as oxygen in medium are changed according to loaded pressure. On the other hand, compression of medium through liquid-phase has been realized in closed system, allowing short-term culture. Therefore, a system for long-term culture under hydrostatic pressure requires medium perfusion and compression through liquid-phase [10, 11].

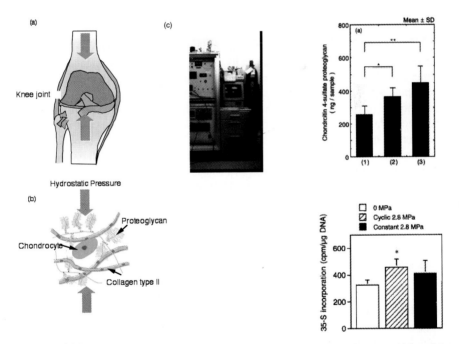

Figure 3. Effects of hydrostatic pressure on matrix production of articular chondrocytes. (a) knee joint, (b) chondrocyte physiologically loaded with hydrostatic pressure, (c) hydrostatic pressure loading system. (d) promotion of chondroitin-4-sulfate and 36-S incorporation of chondrocytes under hydrostatic pressure (Ref. 11 & 12).

Although mechanism how chondrocytes respond to hydrostatic pressure is still unknown, intermittent hydrostatic pressure loading promotes glycosaminoglycans production in three-dimensional culture using collagen sponges [12] (Fig.3). It could be thought for hydrostatic pressure to be one of the essential factors for regenerating cartilage-like tissue in vitro. It is also reasonable to think that physical stimulation is a methodology for regenerating tissues including cartilage, bone and blood vessel in vitro.

4   Approach to Regulate Cell Differentiation and Tissue Regeneration by Means of Spheroid Formation

Cell aggregation is recognized to be essential for cell differentiation in specific tissue such as cartilage at developmental steps. Cell-cell interaction realized in cell aggregation is thought to play a pivotal role in cell differentiation and tissue regeneration. In cell culture in vitro, cell aggregation is known to be significant for regulating cell differentiation and functions in hepatocytes and neuronal cells. The spheroids (cell aggregates) of those cells could be formed on specially treated polymer surfaces or under specific rheological conditions. Those methods could be also adapted to other cells including primary culture cells. For example, fibroblasts, which require oxygen and nutrition supply in higher levels, could form spheroids under adequate rotating flow conditions, having diameters which avoid necrosis at the center of spheroids [13]. The conditions of spheroid formation depend on not only rheological fields but also strength of cell-cell interaction, which changes with biochemical and physical stimulations [14]. Those spheroids of fibroblasts could be adapted for regenerating cultured dermis. There are two kinds of methods: one is to form cultured dermis with only cells and matrices produced by the cells, the other is to form it by using cells and scaffolds such as meshes fabricated with biodegradable polymers, which guarantee mechanical strength and easy handling. The spheroids could be adapted to the latter method, where they are trapped onto the meshes having several hundreds micrometers gaps, and enable prompt tissue regeneration [15].

Cell condensation is detected at developmental steps, and is thought to play a crucial role in chondrocyte differentiation. On the other hand, chondrocytes are known to be dedifferentiated in vitro. Cell condensation is also considered to be essential for regulating differentiation of chondrocytes or maintaining their phenotype. According to this context, cell aggregation is made as a pellet by centrifugating cell suspension, which is frequently used for studying differentiation of mesenchymal stem cells to chondrocytes. However, that pellet culture is limited in basic research, because only one cell aggregate be made by a centrifugation tube. Then, it was found to be able to form cell aggregates of chondrocytes as well as those of fibroblasts by using rotating flows, taking merits of pellet culture and compensating its demerits [16]. Those cell aggregates could be thought to be not only a kind of incubators for cell differentiation but also a kind of micro-tissue elements for tissue regeneration. (Fig. 4) In this sense, cell aggregates are adaptable to diverse kinds of tissue regeneration. Although angiogenesis is indispensable for oxygen and nutrient supply in tissue regeneration, usage of cell aggregates as starting tissue elements shows a possibility to regenerate tissues, avoiding necrosis.

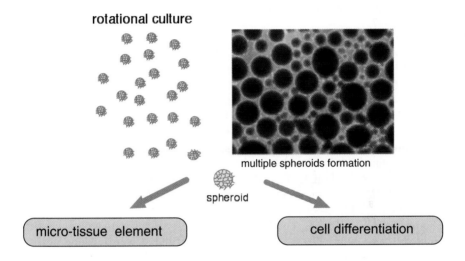

Figure 4. Simultaneous multiple spheroid formation by rotational culture for regulating cell differentiation and formating micro-tissue elements.

5 Summary

There exist huge obstacles to be overcome for realizing not cell transplantation but tissue transplantation, tissue regeneration, as mentioned in the introduction. Even though simultaneous angiogenesis is essential, approaches from scaffolds, and from physical stimulation as well as from biochemical stimulation, are crucial for regulating cell differentiation and regenerating tissues.

Acknowledgements

This work was supported by Grant-in-Aid for Scientific Research on Priority Areas 15086205 from the Ministry of Education, Culture, Sports, Science and Technology of Japan.

References

1. Chen, G., Ushida, T., Tateishi, T., 1999. Fabrication of PLGA-Collagen hybrid sponge, Chem. Lett., 561-562.
2. Chen, G., Ushida, T., Tateishi, T., 2000. A biodegradable hybrid sponge nested with collagen microsponges. J Biomed Mater Res 51 (2), 273-279.
3. Chen, G., Ushida, T., Tateishi, T., 2000. Hybrid biomaterials for tissue engineering: a preparative method for PLA or PLGA-Collagen. Adv. Mater. 12 (6), 455-457.
4. Chen, G., Ushida, T., Tateishi, T., 2000. A hybrid network of synthetic polymer mesh and collagen sponge. J. Chem. Soc. Chem. Comm. 16, 1505-1506.
5. Chen, G., Ushida, T., Tateishi, T., 2001 Poly(DL-lactic-co-glycolic acid) sponge hybridized with collagen microsponges and deposited apatite particulates. J. Biomed. Mater. Res. 57, 8-14.
6. Ushida, T., Furukawa, K., Toita, K. et al, 2002. Three dimensional seeding of chondrocytes encapsulated in collagen gel into PLLA scaffolds. Cell Transplantation 11(5), 489-494.
7. Chen, G., Sato, T., Ushida, T. et al., 2003. Redifferentiation of dedifferentiated bovine chondrocytes when cultured in vitro in a PLGA-collagen hybrid mesh. FEBS Letters 542, 95-99.
8. Furukawa, K., Ushida, T., Toita, K. et al., 2002. Hybrid of gel-cultured smooth muscle cells with PLLA sponge as a scaffold towards blood vessel regeneration. Cell Transplantation 11(5), 475-480.
9. Ochi, K., Chen, G., Ushida, T. et al., 2002. Use of isolated mature osteoblasts in bundance acts as desired-shaped bone regeneration in combination with a modified poly-DL-Lactic-Co-Glycolic acid (PLGA)-collagen sponge. J. Cell Physiol 194, 45-53.
10. Murata, T., Ushida, T., Mizuno, S. et al., 1998. Proteoglycan synthesis by chondrocytes cultured under hydrostatic pressure and perfusion. Material Science & Engineering C 6, 297-300.
11. Mizuno, S., Ushida, T., Tateishi, T. et al., 1998. Effects of physical stimulation on chondrogenesis in vitro. Material Science & Engineering C 6, 301-306.
12. Mizuno, S., Tateishi, T., Ushida, T. et al., 2002. Hydrostatic fluid pressure enhances matrix synthesis and accumulation by bovine chondrocytes in three-dimensional culture. J. Cell Physiol. 193, 319-327.
13. Furukawa, K., Ushida, T., Sakai, Y. et al., 2001. Formation of human fibroblast aggregates (Spheroids) by rotational culture. Cell Transplantation 10, 441-445.
14. Furukawa, K., Ushida, T., Kunii, K. et al., 2001. Effects of hormone and growth factor on formation of fibroblast-aggregates for tissue-engineered skin. Materials and Science & Engineering: C 17 (1-2), 59-62.

15. Furukawa, K., Ushida, T., Sakai, Y. et al., 2001. Tissue-engineered skin using aggregates of normal human skin fibroblasts and biodegradable material. J Artif Organs 4 (4), 353-356.
16. Furukawa, K., Suenaga, H., Toita, K. et al., 2003. Rapid and large-scale formation of chondrocytes aggregates by rotational culture. Cell Transplantation 12(5), 475-479.

A NEW THEORY ON THE LOCALIZATION OF VASCULAR DISEASES

T. KARINO

Laboratory of Biofluid Dynamics, Research Institute for Electronic Science,
Hokkaido University, North 12, West 6, North District, Sapporo 060-0812, Japan
E-mail: karino@bfd.es.hokudai.ac.jp

S. WADA

Physiological Flow Studies Laboratory, Department of Bioengineering and Robotics,
Graduate School of Engineering, Tohoku University,
6-6-01 Aoba-yama, Sendai 980-8579, Japan
E-mail: shigeo@pfsl.mech.tohoku.ac.jp

T. NAIKI

Laboratory of Cellular Cybernetics, Division of Bioengineering and Bioinformatics,
Graduate School of Information Science and Technology, Hokkaido University,
North 14, West 9, North District, Sapporo 060-0814, Japan
E-mail: naiki@cellc.ist.hokudai.ac.jp

A new theory (hypothesis) was proposed to explain the localization of vascular diseases that accompany thickening or thinning of a vessel wall, and to substantiate the hypothesis, the effects of various physical and hemodynamic factors on the concentration of plasma proteins and lipoproteins including low density lipoproteins (LDL) at the luminal surface of an arterial wall were studied both theoretically and experimentally by carrying out computer simulations of the flow and transport of lipoproteins from flowing blood to an arterial wall, and performing mass transfer experiments of LDL and model particles using a monolayer of cultured bovine aortic endothelial cells as a model of an arterial wall. It was found both theoretically and experimentally that due to the presence of a filtration flow of water at the vessel wall, flow-dependent concentration or depletion of LDL occurs at the luminal surface of an artery under physiological conditions, creating regions of high and low LDL concentration at sites of low wall shear rate (stress) and high wall shear rate (stress), respectively. The results strongly suggest that concentration polarization of LDL is responsible for localized genesis and development of atherosclerosis and intimal hyperplasia in regions of low wall shear rate where concentration of LDL builds up, and cerebral aneurysms in regions of high wall shear rate where depletion of LDL occurs.

1 Introduction

It has been shown that, in humans, most of the vascular diseases such as atherosclerosis, cerebral aneurysms, and anastomotic intimal hyperplasia develop preferentially at certain sites in the circulation such as bifurcations, curved segments, and distal to a stenosis of relatively large arteries where flow is locally disturbed by the formation of secondary and recirculation flows. Thus the localization of these vascular diseases has been a subject of hot issue since the early 1970's, and it has been studied by many researchers in the fields of medicine, engineering, and physics with respect to its connection with blood flow. As the results, it has been pointed out that wall shear stress is a key hemodynamic factor playing a causative roll in all the vascular diseases which accompany thickening or thinning of the vessel wall. However, we have a strong suspicion in it for the following reasons.

1. Shear stress is not a substance which can cause the vascular disease directly by accumulation or depletion of itself, but one form of force which can only affect or stimulate the cells constituting a vessel wall to uptake or synthesize or degrade the substances that cause the vascular disease.

2. The results from our flow studies with isolated transparent natural vessels showed that:
   - Atherosclerosis and anastomotic intimal hyperplasia develop preferentially in regions where flow is slow and wall shear rate (shear stress) is low [1, 2].
   - Cerebral aneurysms are formed selectively at sites where blood flow is fast and vessel walls are exposed to high hemodynamic stresses (pressure, tension, and shear stress) [3].

These results indicate that both of the vascular diseases which accompany thickening of the vessel wall (atherosclerosis and anastomotic intimal hyperplasia) and thinning of the vessel wall (saccular cerebral aneurysms) are directly affected by the velocity of blood flow and wall shear rate associated with it, but they are opposite phenomena which occur under extremely low and high velocity, respectively.

In addition to the above reasons, clinical data shows that:

1. Atherosclerosis and anastomotic intimal hyperplasia tend to develop in subjects whose cholesterol concentration in blood is elevated [4-6].

2. Cerebral hemorrhagic disorders tend to occur in subjects whose cholesterol concentration in blood is low [7].

3. In autologous vein grafts and artificial grafts implanted in the arterial system, even if their diameters are adjusted to those of host arteries and the luminal surfaces are exposed to wall shear stress which is the same level as that of host arteries, intimal hyperplasia still develops in the grafts [8].

From all these facts and data, we considered that the flow velocity of blood and cholesterol (as a nutriment) are involved in all the vascular diseases, and the concentration of cholesterol in blood and its transport to vessel walls play important roles in the localized pathogenesis and development of the vascular diseases mentioned above.

To this end, we developed our own new theory (hypothesis) to explain the pathogenesis and localization of the vascular diseases.

2 Our Theory (hypothesis)
" Flow-dependent concentration polarization theory"

Due to a semi-permeable nature of a vascular endothelium which allows the passage of water and water-dissolved ions but not macromolecules such as plasma proteins and lipoproteins, flow-dependent concentration or depletion of plasma proteins and lipoproteins occurs at a blood-endothelium boundary as shown in Figure 1, and this affects the transport of lipoproteins which carry an important nutriment 'cholesterol' from flowing blood to an arterial wall, leading to localized genesis and development of atherosclerosis and intimal hyperplasia in regions of low wall shear rate where concentration of these macromolecules builds up, and cerebral aneurysms in regions of high wall shear rate where depletion of macromolecules occurs.

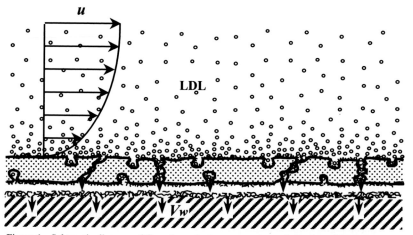

Figure 1. Schematic diagram of flow-dependent concentration polarization of LDL. Due to the presence of a filtration flow of water at a vessel wall, there occurs concentration or depletion of LDL at the luminal surface of an artery, and depending on the flow rate, a constant equilibrium concentration is established. EC: endothelial cell, LDL: low density lipoprotein, Vw: filtration velocity of water (From Wada and Karino [9]).

3 Theoretical Study

3.1 Method

The effects of various physical and fluid mechanical factors on the concentration of low density lipoproteins (LDL), known to be a substance which cause aterosclerosis, at a blood–endothelium boundary in various segments of arteries were studied by solving equations of motion, continuity, and mass transport under the assumptions that blood is an incompressible, homogeneous Newtonian fluid, the vessel wall is permeable to water (plasma), and blood flow at the inlet of the vessel is a steady laminar flow with a parabolic velocity distribution.

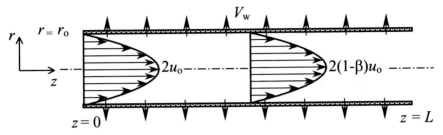

Figure 2. Schematic representation of a straight artery and definitions of symbols used (From Wada and Karino [9]).

Steady state mass transport of macromolecules such as LDL in blood flowing through a straight artery can be described by

$$u\frac{\partial C}{\partial z} + v\frac{\partial C}{\partial r} - D\left(\frac{1}{r}\frac{\partial C}{\partial r} + \frac{\partial^2 C}{\partial r^2}\right) = 0$$

where C is the concentration of the macromolecule, D is the diffusivity of the macromolecule, z and r are the respective axial and radial coordinates, and u and v are the velocity components of the fluid in the axial and radial direction, respectively.

The boundary conditions applied to solve the above equation are:

$$C = C_o \qquad\qquad \text{at} \quad z = 0$$

$$\frac{\partial C}{\partial r} = 0 \qquad\qquad \text{at} \quad r = 0$$

$$V_w C_w - D\frac{\partial C}{\partial r}\bigg|_{r=r_o} = K C_w \qquad\qquad \text{at} \quad r = r_0$$

where r_0 is the radius of the vessel, C_w is the surface concentration of LDL at a blood–endothelium boundary, Vw is the filtration velocity of water, and K is the overall mass transfer coefficient of LDL at the vessel wall.

3.2   Procedures for computational analyses

Numerical solution for LDL concentration, C, was obtained by a finite element method. To do so, solution domain was first discretized using two-dimensional, bilinear, quadrilateral elements, and the element equation was formulated by a conventional Galerkin-weighted residual method.

3.3 Results

3.3.1 Straight artery [9]

The axial and radial velocity of blood and the concentration of lipoproteins in blood flowing through straight arteries in a steady fashion were calculated by assuming the filtration velocity of water at the vessel wall, $V_w = 4 \times 10^{-6}$ cm/s [10], the overall mass transfer coefficient of LDL at the vessel wall, $K = 2 \times 10^{-8}$ cm/s [11], and the diffusivity of LDL, $D = 5 \times 10^{-8}$ cm^2/s [12]. Figure 3 shows the distribution of the concentration of various lipoproteins including LDL in the radial direction at an axial location $z/d_o = 20$ (20-diameter downstream from the entrance of the vessel) calculated for a 0.6-cm inner diameter artery at Reynolds number, Re = 100. Here the local concentration of lipoproteins and radial distance from the vessel axis were normalized by the concentration at the entrance, C_o, and the radius of the artery, r_o, respectively. As evident from the figure, it was found first that due to the presence of a filtration flow of water at the vessel wall, concentration polarization of lipoproteins occurred at a blood–endothelium boundary where fluid axial velocity was small. The degree of polarization was much greater for large lipoproteins such as LDL ($D = 5 \times 10^{-8}$ cm^2/s) and VLDL ($D = 1 \times 10^{-8}$ cm^2/s) than albumin ($D = 5 \times 10^{-7}$ cm^2/s).

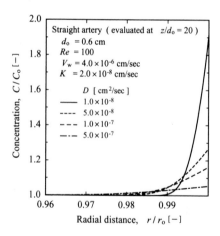

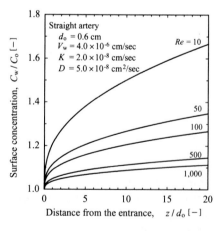

Figure 3. The effect of the diffusivity of lipoproteins on concentration profile near the vessel wall (From Wada and Karino [9]).

Figure 4. The change in surface concentration of lipoproteins along the artery, and the effect of Reynolds number on it (From Wada and Karino [9]).

It was also found that the surface concentration of LDL increased non-linearly with increasing the axial distance from the entrance of the artery (z/d_o), and it was greatly affected by the flow rate (Re) as shown in Figure 4. Figure 5 shows the relationship between the surface concentration of LDL and wall shear rate or Re. As it is evident from the figure, concentration polarization of LDL occurred most prominently under the condition of low Re and low wall shear rate (shear stress).

The effect of the filtration velocity of water at the vessel wall on surface concentration of LDL was also studied. It was found that, as shown in Figure 6, the effect of water filtration was very large at low flow rates (Re), but it became very small at high flow rates (Re).

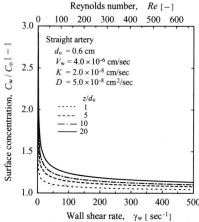

Figure 5. The relationship between the surface concentration of LDL and wall shear rate or Re (From Wada and Karino [9]).

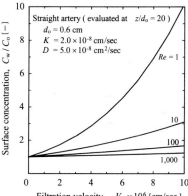

Figure 6. The effect of water filtration velocity on surface concentration of LDL (From Wada and Karino [9]).

3.3.2 Endothelium [13]

A similar analysis was carried out also for the case of an endothelium by simulating its luminal surface with a sinusoidal one [13]. It was found that, as shown in Figure 7, the surface concentration of LDL increased in going distally and took locally high and low values at the valleys and hills of the endothelium corresponding to the sites where wall shear rate (shear stress) was low and high, respectively, and it increased 4.5% even in a short distance of 154 μm (corresponding to the length of 5 cells). The results indicate that concentration polarization of LDL certainly occurs at the luminal surface of the endothelium even though the filtration velocity of water at the endothelium is extremely small and the flow in the vicinity of the endothelium is disturbed microscopically by the presence of the hills and valleys of the monolayer of endothelial cells.

$$u_o = 9.23 \times 10^{-2} \text{ cm/s}$$
$$D = 5.0 \times 10^{-8} \text{ cm}^2/\text{s}$$
$$V_w = 4.0 \times 10^{-6} \text{ cm/s}$$
$$K = 2.0 \times 10^{-8} \text{ cm/s}$$

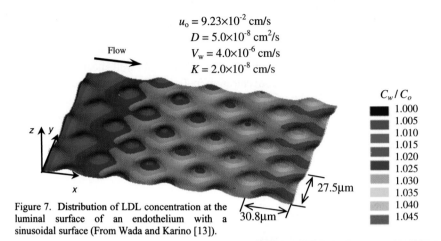

C_w / C_o

1.000
1.005
1.010
1.015
1.020
1.025
1.030
1.035
1.040
1.045

Figure 7. Distribution of LDL concentration at the luminal surface of an endothelium with a sinusoidal surface (From Wada and Karino [13]).

3.3.3 Curved segments [14]

Theoretical study on the effects of various physical and hemodynamic factors on transport of LDL from flowing blood to the wall of an artery with a multiple bend was carried out by means of computer simulation under the condition of a steady flow. Here the shape of the blood vessel was obtained from a photograph of a transparent human right coronary artery shown in Figure 8A and previously used in the study of Asakura and Karino [1]. As evident from the thickness of the wall, the vessel contained an atherosclerotic intimal thickening distal to the apex of the inner wall of the acute second bend where a slow recirculation flow was formed and wall shear rate (shear stress) was low [1]. It was found that due to a semipermeable nature of the vessel wall to plasma, flow-dependent concentration polarization of LDL occurred at the luminal surface of the vessel, creating a region of high LDL concentration distal to the apex of the inner wall of each bend where the flow was locally disturbed by the formation of

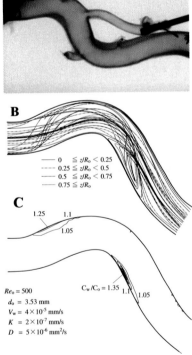

Figure 8. A photograph of a transparent human right coronary artery (A) and obtained flow patterns (B) and surface concentration of LDL (C) (From Wada and Karino [14]).

secondary and recirculation flows and where wall shear rates (shear stresses) were low. The highest surface concentration of LDL occurred distal to the acute second bend where atherosclerotic intimal thickening developed. At $Re_o = 500$, the values calculated using estimated diffusivity of LDL in whole blood and plasma were respectively 35.1% and 15.6% higher than that in the bulk flow.

3.3.4 End-to-end anastomosed artery [15]

Computer simulation of blood flow and transport of LDL from flowing blood to an arterial wall was carried out also for the case of end-to-end anastomosed arteries under the conditions of a steady flow. Here the shapes of the blood vessels were obtained from tracings of transparent dog femoral arteries containing a 45°-cut end-to-end anastomosis that were used in the flow study of Ishibashi et al. [2]. It was found that due to a semipermeable nature of an arterial wall to plasma, concentration polarization of LDL occurred also in anastomosed vessels, but only in vessels with a stenosis in which a slow recirculation flow occurred distal to an anastomotic junction. In the particular vessel shown in Figure 9A, the surface concentration of LDL was locally elevated by more than 10% in a restricted region distal to the stenosis where a recirculation flow was formed, whereas in other regions, it increased only 5% or less. The highest value of the surface concentration, which was found just downstream of the apex of the stenosis, was 1.19, and the location corresponded well to the site where intimal thickening developed.

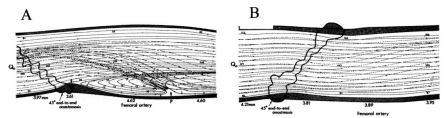

Figure 9. Tracings of 45°-cut and end-to-end anastomosed dog femoral arteries from which computational models were constructed. A: Anastomosed artery containing a moderate stenosis and intimal thickening on the inferior wall. B: Anastomosed artery containing neither a stenosis nor intimal thickening (From Ishibashi et al. [2]).

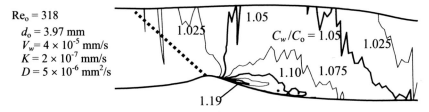

Figure 10. Contour map of LDL concentration at the luminal surface of an anastomosed artery containing a moderate stenosis as observed normal to the bisector plane of the vessel. Contor lines were drawn at an interval of 2.5% change in normalized surface concentration of LDL (From Wada et al. [15]).

4 Experimental Study

4.1 Method

To substantiate the findings of the theoretical study described above experimentally, the effect of a steady shear flow on concentration polarization of macromolecules was studied using a cultured bovine aortic endothelial cell (BAEC) monolayer which served as a model of an endothelium of an artery or an implanted vascular graft and a suspension of plasma proteins and lipoproteins.

The study was carried out by flowing a cell culture medium containing fetal calf serum or bovine plasma lipoproteins in steady flow through a parallel-plate flow cell shown in Figure 11 in which a cultured BAEC monolayer was installed as a part of the parallel-plate under physiological ranges of wall shear rate and water filtration velocity at the BAEC monolayer, and measuring water filtration velocity at the BAEC monolayer which varied secondarily with the change in the concentration of plasma protein particles at the luminal surface of the BAEC monolayer.

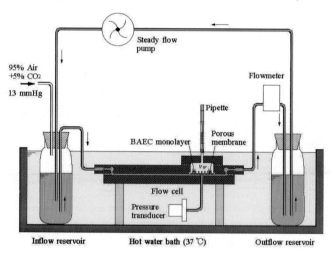

Figure 11. Schematic diagram of the steady flow perfusion system used (From Naiki et al. [16]).

4.2 Results

With perfusates containing serum or lipoproteins, water filtration velocity varied as a function of wall shear rate, and it took the lowest value at $\gamma_w = 0$. The above phenomenon was observed only with perfusates containing lipoproteins. It did not occur with a pure cell culture medium nor a culture medium containing albumin. As shown in Figure 12, water filtration velocity increased or decreased as flow rate increased or decreased from an arbitrarily set non-zero value, indicating that surface concentration of lipoprotein particles varied as a direct function of wall shear rate (flow rate). Almost the same result was obtained with a perfusate containing serum at 20% by volume and a perfusate which contained lipoproteins in an amount equivalent to that contained in a perfusate containing serum at 20% by volume as shown in Figure 13, indicating that lipoproteins are main contributors to this phenomenon.

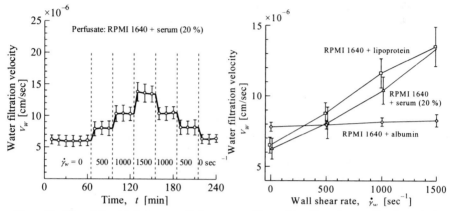

Figure 12. The relationship between wall shear rate and water filtration velocity at the vessel wall which occurs as a result of the change in surface concentration of lipoproteins (From Naiki et al. [16]).

Figure 13. Plot of water filtration velocity against wall shear rate, showing the effect of particle size of lipoproteins on filtration velocity (From Naiki et al. [16]).

5   Concluding Remarks

Through a computer-aided simulation of the flow and transport of LDL in blood flowing through various segments of arteries, we first found that flow-dependent concentration polarization of lipoproteins occurs at the luminal surface of the arteries under physiological conditions. Then by carrying out flow experiments using monolayers of cultured bovine aortic endothelial cell as a model of a vascular endothelium and suspensions of lipoproteins as a model of blood, we experimentally confirmed that the predicted phenomenon certainly occurs at the luminal surface of the endothelium.

The results from both the theoretical and experimental studies showed that the surface concentration of LDL increases with increasing water filtration velocity. If we consider a clinical case of chronic hypertension in the light of the above finding, it is very likely that, in patients suffering from chronic hypertension, due to increased filtration velocity of plasma at the vessel wall, surface concentration of

LDL is chronically elevated, and this is, in turn, promoting the pathogenesis and progression of atherosclerosis. This may account for the clinical finding that atherosclerotic lesions develop preferentially in hypertensive subjects. The importance of our "Concentration polarization theory" is that, with this theory, we can explain not only the effect of flow patterns on surface concentration of LDL but also how the surface concentration of LDL is affected by the concentration of cholesterol in blood and blood pressure that are the two major factors recognized as risk factors of atherosclerosis. As far as we know, this is the only theory which predicts the existence of spatial heterogeneity in LDL surface concentration prior to the development of vascular diseases such as atherosclerosis, anastomotic intimal hyperplasia, and cerebral aneurysms which accompany thickening and thinning of vessel walls. The finding from our theoretical and experimental studies strongly support our hypothesis that flow-dependent concentration polarization of LDL (a carrier of "cholesterol" that is an important component of a cell membrane and thus indispensable to the cells forming the vessel wall for their growth, division, and proliferation) is responsible for localized genesis and development of various vascular diseases mentioned above in man by either locally elevating or lowering the surface concentration of LDL, thus augmenting or reducing their uptake by the cells existing at such sites.

We are still working on this problem, and now investigating the effects of a shear flow and water filtration on surface concentration of LDL at the vessel wall and the uptake of LDL by the cells forming the vessel wall by using a co-culture of endothelial cells and smooth muscle cells as a realistic model of an arterial wall. If we could prove that flow-dependent concentration or depletion of lipoproteins is certainly occurring in real arteries in vivo by carrying out further detailed studies, it might become possible to explain the pathogenesis and localization of most of the vascular diseases that accompany a change in inner diameter of the vessel and remodeling of vessel wall in terms of an excessive or insufficient supply of an important nutriment "cholesterol" from flowing blood to the vessel wall. It might become also possible to treat intimal hyperplasia and atherosclerosis by locally lowering water permeability at the vessel wall by topically applying certain drugs and carrying out some simple surgical procedures.

Acknowledgements

This work was supported by Grant-in-Aid for Scientific Research 15086201 from the Ministry of Education, Science, Sports and Culture of Japan and partly by a Grant-in-Aid for Scientific Research 15300150 from Japan Society for the Promotion of Science (JSPS).

References

1. Asakura, T., Karino, T., 1990. Flow patterns and spatial distribution of atherosclerotic lesions in human coronary arteries. Circ. Res. 66, 1045-1066.
2. Ishibashi, H.,Sunamura, M.,Karino, T., 1995. Flow patterns and preferred sites of intimal thickening in end-to-end anastomosed vessels. Surgery 117, 409-420.
3. Karino, T., 1986. Microscopic structure of disturbed flows in the arterial and venous systems, and its implication in the localization of vascular diseases. Inter. Angiol. 4, 297-313.
4. Ross, R., Harker, L., 1976. Hyperlipidemia and atherosclerosis. Chronic hyperlipidemia initiates and maintains lesions by endothelial cell desquamation and lipid accumulation. Science 193, 1094-1100.
5. Small, D.M., 1988. Progression and regression of atherosclerotic lesions. Insights from lipid physical biochemistry. Arteriosclerosis (Dallas) 8, 103-129.
6. Baumann, D.S., Doblas, M., Daugherty, A., Sicard, G., Schonfeld, G., 1994. The role of cholesterol accumulation in prosthetic vascular graft anastomotic intimal hyperplasia. J. Vasc. Surg. 19, 435-445.
7. Puddey, I.B., 1996. Low serum cholesterol and the risk of cerebral haemorrhage. Atherosclerosis 119, 1-6.
8. Mii, S., Okadome, K., Onohara, T., Yamamura, S., Sugimachi, K., 1990. Intimal thickening and permeability of arterial autogenous vein graft in a canine poor-runoff model: transmission electron microscopic evidence. Surgery 108, 81-89.
9. Wada, S., Karino T., 1999. Theoretical study on flow-dependent concentration polarization of low density lipoproteins at the luminal surface of a straight artery. Biorheology 36, 207-223.
10. Wilens, S.L., Mccluskey, R.T., 1952. The comparative filtration properties of excised arteries and veins. Am. J. Med. Sci. 224, 540-547.
11. Truskey, G.A., Roberts,W.L., Herrmann, R.A., Malinauskas, R.A., 1992. Measurement of endothelial permeability to ^{125}I-low-density lipoproteins in rabbit arteries by use of en face preparations. Circ. Res. 71, 883-897.
12. Back, L.H., 1975. Theoretical investigation of mass transport to arterial walls in various blood flow regions. I. Flow field and lipoproteins transport. Mathematical Biosciences 27, 231-262.
13. Wada, S., Karino, T., 2002. Prediction of LDL concentration at the luminal surface of a vascular endothelium. Biorheology 39, 331-336.
14. Wada, S., Karino, T., 2002. Theoretical prediction of low-density lipoproteins concentration at the luminal surface of an artery with a multiple bend. Ann. Biomed. Eng. 30, 778-791.
15. Wada, S., Koujiya, M., Karino, T., 2002. Theoretical study of the effect of local flow disturbances on the concentration of low density lipoproteins at the luminal surface of end-to-end anastomosed vessels. Med. Biol. Eng. Compt. 40, 576-587.

16. Naiki, T., Sugiyama, H., Tashiro, R., Karino, T., 1999. Flow-dependent concentration polarization of plasma proteins at the luminal surface of a cultured endothelial cell monolayer. Biorheology 36, 225-241.

AUTOMORPHOGENESIS OF LOAD BEARING FIBROUS TISSUES: GENERATION OF TENSILE STRESS, CELL ALIGNMENT, AND MATRIX DEFORMATION BY FIBROBLASTS

K. TAKAKUDA

*Tokyo Medical and Dental University, 2-3-10 Kanda-Surugadai, Chiyoda-ku,
Tokyo 101-0062, Japan
E-mail: takakuda.mech@tmd.ac.jp*

Tensile stresses generated by fibroblasts cultured in collagen gel were investigated. Specimens of thin collagen gel membrane, within which cells were proliferating, were examined under various initial and boundary conditions. Generation of tensile stress, cell alignment, and matrix deformation were recorded. It was demonstrated that fibroblasts generate tension, change their orientation along tensile direction, and create structures of collagen fibers. A hypothetical mechanism for such automorphogenesis was proposed, that is, fibroblasts generate tension and make tense collagen fibers, and then cells stretch themselves along the tense fibers and increase tension in this direction. Thus this mechanism works as positive feedback that enable cells to make load bearing fibrous tissues with collagen fibers.

1 Introduction

Soft fibrous connective tissues such as tendons and ligaments in our body are mechanical components bearing tensile loads. These tissues remodel themselves to adapt to the mechanical loadings. They undergo hypertrophy if they are subjected to excessive loads, and atrophy to reduced loads [1, 2]. The mechanism how such remodeling realized is hardly elucidated, but fibroblasts are the dominant cell population within these tissues and are believed to recognize and respond appropriately to the mechanical environment around the tissues.

Actually some mechanical activities of cells were observed in culture system, as the contraction of collagen gel within which fibroblasts were proliferating [3], and the alignment of fibroblasts along the lines connecting two particles to which collagen fibers were attached [4-6]. These observations made us believe that fibroblasts express mechanical functions, and these functions could be analyzed in detail with the use of *in vitro* models. Hence we have been developing experimental models in which fibroblasts were cultured in thin collagen gel membranes and subjected to various initial and boundary conditions [7-9]. These models have been proved quite convenient for mechanical manipulations and microscopic observations. Utilizing these models here we would demonstrate that fibroblasts generate tension, change their orientation along tensile direction, and create structures of collagen fibers. Furthermore, based on these observations, we would discuss automorphogenesis of fibrous connective tissues from the mechanical point of view.

2   Materials and Methods

The fibroblasts-like cells were obtained by the explants method from synovial membranes of knee joints of Japanese white rabbits. They were cultured in modified MEM with nucleosides (JRH Bioscience) supplemented by 60 µg/ml Kanamycin and 10 % calf serum, and kept in an incubator at 37 degrees Celsius and 5 % CO_2 environment. They were subcultured through the dispersion of cells by trypsin/EDTA solution. For the experiments, the cells from 1st to 5th subculture were used.

A typical specimen is illustrated in Figure 1. A collagen gel membrane of 10x10 mm was kept on a stainless steel mesh (#80) with a square hole of 5x5 mm. This specimen was made by pouring a mixture of 8 parts of collagen acidic solution (CELL MATRIX type I-A, NITTA Gelatin), 1 part of x10 solution of MEM (Nissui), and 1 part of buffered solution (4.77g HEPES/100ml 0.08N NaOH solution) into a silicone rubber mold in which the stainless steel had been set previously, and keeping it in a incubator for 15 min for gelation. The thickness of the specimens was about 0.5 mm although it could not be set exactly since the upper surface of the mold was left open.

Many variant specimens were made, those having holes or straight cuts within the gel membrane. The fixed boundary conditions were realized on the boundaries of the specimens where they were supported by the stainless steel wire mesh. The stress free boundary conditions were realized on the free boundaries of the specimens where a cut was introduced in the specimens by a surgical knife or where a hole was made at the preparation of the specimens.

The distribution of cells in a specimen was also varied. In the case of a specimen with uniformly distributing cells was required, cells were dispersed in the neutralized collagen solution before gelation. In the case of a specimen with non-uniformly distributing cells was required, a monolayer patch of cells was retrieved from a culture dish and the mass of cells (*ca.* 1 mm^3) was attached on the stainless steel mesh near the edge of the opening just before pouring the neutralized collagen solution. Outgrown cells proliferated and migrated from the mass of the cells, and we got specimens with non-uniformly distributing cells.

After gelation, the specimens were placed in plastic petri dishes (60 mm diameter) filled with the medium and kept in the incubator. The medium was changed every 3 or 4

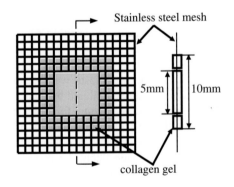

Figure 1. Collagen gel specimen.

days. The specimens were observed by a phase-contrast microscope (DIAPHOTO TMD, Nikon) equipped with an incubating box. The specimens were inspected everyday and time-lapse videos were recorded at occasions of interests.

3 Results

3.1 Generation of tensile stresses by fibroblasts

Setting a small mass of the cells near an inner edge of the steel mesh at the making of the specimen, we could make the cells proliferate and migrate in the radial direction from the mass of the cells into the gel membrane [8]. When the front of the cells reached to the central portion of the specimen, we made a cut in the gel membrane by a surgical knife as illustrated in Figure 2(a). Then the cut opened at once as Figures 2 (b) and (c) show. Furthermore, as shown in Figure 2(d), the gel of the upper side above the cut in the figure, within which many cells could be observed, showed progressive deformation with time. On the other hand, the gel of the lower side below the cut, within which no cells could be observed, did not show any deformation.

The observation indicates that the tensile stress field was generated in the membranes. When we cut the gel membrane by a knife, the residual stresses were released and the cut opened. On the other hand, in the specimens which did not

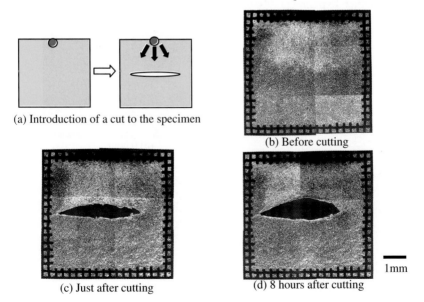

(a) Introduction of a cut to the specimen

(b) Before cutting

(c) Just after cutting

(d) 8 hours after cutting

1mm

Figure 2. Release of internal stresses generated by cells.

containing the cells, this opening phenomena described above could not be observed. These facts indicate that the cells generated the tensile stresses.

3.2 Alignment of fibroblasts in the direction of the tensile stress

In order to investigate the behavior of cells in the stressed gel, a specimen was made in which the cells were proliferating and migrating from a mass of the cells again, and two knife-cuts were introduced along the both sides of specimen as shown in Figure 3(a). In such a specimen, we could make the cells proliferating in the tensile stress field [9]. In fact, when the cells generated the tensile stresses, since only the upper and the lower boundary of the specimen were fixed, the stress should be tensile stress between the upper and the lower boundaries. The behavior of the cells was recorded with the time-lapse video system, at the time when frontal cells had reached to the central portion of the specimen. Typical behavior of the cells is shown in Figures 3(b)-(d). The time indicated in photos was measured from the beginning of the recording. The cells were apparently lying on the surface of gel membrane. Most of cells took the elongated form in the tensile direction. Cells took the rounded form just before the mitosis, and then newborn daughter cells recovered elongated form in the tensile direction immediately. This re-elongation process was very rapid. The cell migrations in the tensile direction were also observed.

In another experiment for the cell alignment, we utilized the same kind of

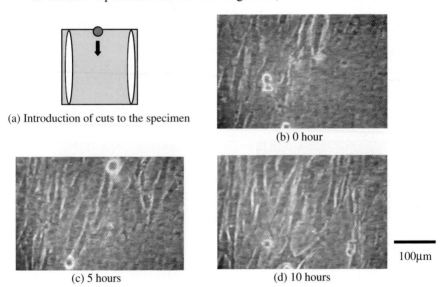

(a) Introduction of cuts to the specimen

(b) 0 hour

(c) 5 hours

(d) 10 hours

100μm

Figure 3. Cell alignment in the tensile stress direction.

specimen as in the section 3.1 and recorded the behavior of cells with the time-lapse video system. When frontal cells had reached to the central portion of the specimen, the cells took the elongated shape in vertical direction. At this time, we introduced a knife-cut at the central portion of the specimen in horizontal direction as before. During about 10 minutes after the introduction of the knife-cut, the cells took the rounded forms as shown in Figure 4(b). Then as time passed, they took the elongated shapes again but in horizontal direction in the figure as Figures 4(c) and (d) show. Whole process had completed within an hour. The cells elongated themselves as if they knew the direction to elongate, although the details were not clear in our phase-contrast microscopy. It should be noted that when the number of cells was so small in the specimens, this reorientation phenomena described above could not be observed.

3.3   Matrix deformation by fibroblasts

In order to investigate the matrix deformation caused by the cells, specimens with a small hole with uniformly distributing cells were examined [7, 8]. Figure 5 shows the experimental results of the specimen with a hole of 0.5 mm diameter and cells with density of 0.2×10^6/ml. According to the result of the section 3.1, the

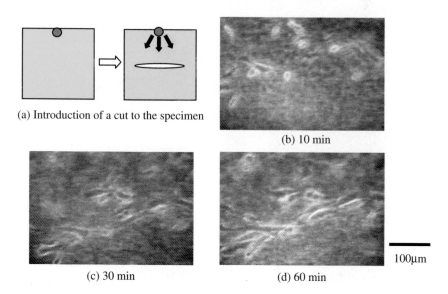

(a) Introduction of a cut to the specimen

(b) 10 min

(c) 30 min   (d) 60 min

100μm

Figure 4. Reorientation of cells toward the tensile stress direction.

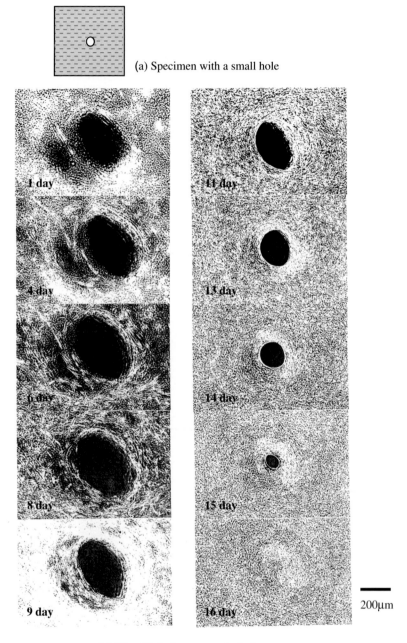

(a) Specimen with a small hole

200μm

(b) Magnified views around a hole

Figure 5. Cell alignment and matrix deformation around a small hole.

tensile stress fields were generated within the collagen gel membrane. Around the hole, the stress field might be such one that it was tensile in the circumferential direction and stress-free in the radial direction. In the experiment, the cells aligned circumferentially near the edge of the hole, demonstrating clearly that the cells oriented themselves along the direction of the tensile stress. As time passed, the diameter of the hole decreased and the hole finally disappeared. This seemingly resulted from the gel contraction caused by the circumferential tensile stress.

Another example of the matrix deformation caused by the cells over a long period of time is shown in Figure 6. Initially, the cells were distributing uniformly with a density of 0.2×10^6/ml, and two cuts were introduced along both sides of the specimen just after the specimen was prepared. The cells aligned in a direction parallel to the cuts, *i.e.,* along the tensile direction. Under the tensile stress generated by the cells, the stretching of specimen proceeded with time and finally the specimen ruptured spontaneously.

4   Discussion

4.1   Generation of tensile stresses by fibroblasts

It is well known that if a specimen of collagen gel containing fibroblasts within it is freely drifting in the medium, it will contract as cells proliferate [3]. This

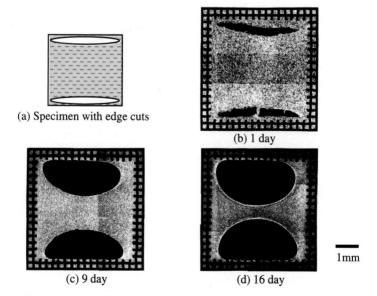

(a) Specimen with edge cuts

(b) 1 day

(c) 9 day

(d) 16 day

1mm

Figure 6. Matrix deformation by tensile force generated by cells.

contraction is believed to be due to the mechanical function of the cells, *i.e.* the exertion of contracting forces. Contrastingly, if the boundary of the specimen is attached to surrounding tissues, a situation that may arise most frequently *in vivo*, tensile stresses should be generated in the collagen gel as the cells contract. Although these stresses cannot be observed directly, the existence of them is easily verified by the technique demonstrated here. The result shown in 3.1 is a clear evidence for the generation of tensile stresses by the cells. The introduction of knife-cut release the generated internal stresses in the gel and the cut opened spontaneously. Progressive shape change of the gel also demonstrates that the cells continued to generate tensile stresses.

4.2   Fibroblasts alignment along the tensile stress direction

In the first experiment shown in the sectoin 3.2, we observed that the round-formed cells just after the mitosis found the direction of the tensile stress and got elongated shape aligned in this direction. This phenomenon suggests that the cells adhered to the collagen fibers and stretched along the fibers. The fact that the cells just after the mitosis so quickly took elongated forms may be explained by the existence of the well-developed orientation of the collagen fibers in the gel. Similarly, in the second experiment in the section 3.2, the cells must be adhered to the collagen fibers in the gel before the introduction of the cut since they exhibit elongated forms. After cutting, the collagen fibers lost their tension. It may correspond to the observation that the cells got the rounded form after the cutting. It looks that the cells could not keep their elongated form with the collagen fibers that lost tension. The cells seemed to lost adhesion to the fibers previously adhered, and then begin to re-adhere to the other fibers. Our results shows that these fibers should be horizontal in the figure and possibly be stretched in this direction. The facts that this process took only an hour and the cells did not seem to looking for the directions to elongate suggests that the cells would recognize the tense collagen fibers. We believe that the cells require some stable anchors to adhere and want to stretch themselves over them. The rigid anchors for the cell adhesion are required, and in the case of the soft materials such as the collagen fibers, they must be in tension to be rigid enough for the cells. Although interactions between the cytoskeltons and the extracellular collagen fibers are believed to be responsible for these phenomena, the detail of the process cannot be elucidated in these observations. Further investigations are required to know how the interaction mechanism works.

4.3   Matrix deformation by fibroblasts and automorhogenesis

From the results shown in the section 3.3, we found that the populations of the cells could generate sufficient force for the induction of the matrix deformations. Hence if randomly orientated cells exert contraction force among them, and if the direction of tensile stress is restricted as in the portion near the free surface of the gel specimen, the collagen fibers are stretched and change their orientation to align in the direction of the tensile stress that may be the direction parallel to the free surface of the specimen. In this case, the cells would adhere to the aligned collagen fibers and they would generate the contracting force along the direction of the collagen fibers. Thus the positive feedback loop between the orientation of the tensile stress and that of the collagen fibers is realized, and the randomly orientated cells in the beginning will align themselves in the tensile direction as they proliferate.

Unfortunately, our experiments presented here also elucidated that the collagen matrix utilized for the cell culture experiments was so weak that it would rupture by the sole application of the forces generated by the cells within the gel itself. Not only the self-orientation mechanism introduced here but also the strengthening mechanism for the extracellular matrix is necessary to establish the appropriate hypothesis for the automorhogenesis of the load bearing fibrous tissues, and it would be presented in our other articles.

Acknowledgements

This work was supported by Grant-in-Aid for Scientific Research on Priority Areas 15086206 from the Ministry of Education, Culture, Sports, Science and Technology of Japan.

References

1. Noyes, F.R., 1977. Functional properties of knee ligaments and alternations induced by immobilization. Clin. Orthop. 123, 210-242.
2. Takakuda, K., Fujii, S., et al., 1993. Mechanical problems in the reconstruction of anterior cruciate ligaments (mechanical compatibility between living tissues and artificial materials). JSME Int J Ser A. 36, 327-332.
3. Bell, E., Ivarsson, B., Merrill, C., 1979. Production of a tissue-like structure by Contraction of Collagen Lattices by Human Fibroblasts of Different proliferative potential *in vitro*. Proc Natl Acad Sci USA. 76, 1274-1278.
4. Bellows, C.G., Melcher, A.H., Brunette, D.M., 1980. Orientation of calvaria and periodontal ligament cells *in vitro* by pairs of demineralized dentine particles. J Cell Sci. 44, 59-73.
5. Bellows, C.G., Melcher, A.H., Aubin, J.E., 1981. Contraction and organization of collagen gels by cells cultured from periodontal ligament, gingiva and bone suggest functional differences between cell types. J. Cell Sci. 50, 299-314.

6. Bellows, C.G., Melcher, A.H., Aubin, J.E., 1982. Association between tension and orientation of periodontal ligament fibroblasts and exogenous collagen fibers in collagen gels *in vitro*, J. Cell Sci. 58,125-138.
7. Takakuda, K., Miyairi, H., 1995. Structures made by fibroblasts *in vitro*. Rep Inst Med Dent Engng. 29, 117-124.
8. Takakuda, K., Miyairi, H., 1996. Tension induced by fibroblasts and automorphogenesis. (in Japanese) Trans. Jpn. Soc. Mech. Engng. 62, 800-807.
9. Takakuda, K., Miyairi, H., 1996. Tensile behavior of fibroblasts cultured in collagen Gel. Biomaterials. 17, 1393-1397.

IV.   COMPUTATIONAL BIOMECHANICS

NOTE ON ANISOTROPIC PROPERTIES OF CANCELLOUS BONE AND TRABECULAE: ELASTICITY AND HARDNESS

M. TANAKA, T. MATSUMOTO AND M. IHARA

Department of Mechanical Science and Bioengineering, Graduate School of Engineering Science, Osaka University, Machikaneyama 1-3, Toyonaka, Osaka 560-8531, Japan
E-mail: tanaka@me.es.osaka-u.ac.jp

M. TODOH

Division of Mechanical Science, Graduate School of Engineering, Hokkaido University, Kita13 Nishi8, Kita-ku, Sapporo 060-8628, Japan

Trabecular architecture is the microstructure of cancellous bone, and it is important to understand the relation between the mechanical properties of trabecula level and those of cancellous level. This article describes the anisotropy of mechanical and structural properties of the cancellous bone observed by compression test and mean intercept length analysis, and the anisotropy of mechanical properties of trabeculae observed by bending test and hardness test of single trabecula. The former anisotropy is considered from the viewpoint of the latter anisotropy, that is the dependency of trabecula properties on the direction of trabecula in the cancellous bone. Findings have suggested collectively a possibility that the anisotropic mechanical properties of cancellous bone are resultant of anisotropy in trabecular structure and trabecula properties.

1 Introduction

Cancellous bone is made of beam- or plate-like trabeculae that construct the trabecular architecture [1]. The mechanical properties as cancellous bone are therefore governed by the mechanical properties of structural elements and the organization of trabecular architecture. It has been known that the mechanical anisotropy as the cancellous bone is coming from the anisotropy of trabecular architecture partially dependent on its mechanical condition [2]. Multiscale observation is inevitable in order to understand the mechanical properties of cancellous bone as the whole. In fact, efforts have been devoted for the investigations concerning the mechanical properties as the cancellous bone through conventional compression test [3, 4], the structural properties as the cancellous bone by means of fabric analysis based on imaging [5, 6], the mechanical properties of trabeculae through tensile/bending tests or indentation test [7, 8] and so on.

This article describes the relation between the anisotropic elastic properties as cancellous bone and the elastic and hardness properties of trabecular elements in conjunction with the structural anisotropy of trabecular architecture. For the cancellous bone scale, the conventional compression test and the mean intercept length analysis of micro CT images are conducted, and the bending test and Vickers hardness test are done fro the mechanical properties of single trabecula scale.

2 Methods

2.1 Specimens

This study uses the cancellous bone of bovine femur as the test specimen. The fresh femurs of young Japanese black cattle are purchased from the meat market and stored at forty degrees below the freezing point until experiment. In order to identify the orientation of the specimen, the x-axis of coordinate system is assigned to the anterior direction, and the z-axis is to the proximal direction of the bone shaft. The y-axis represents the mediolateral direction so that the coordinate system to be right-handed.

Four mechanical/structural tests are conducted in this study: (1) compression test of cancellous bone, (2) mean intercept length analysis of cancellous bone, (3) bending test of single trabecula, and (4) hardness test of single trabecula. The femur is bi-sectioned by sagittal x-z plane on a band saw machine, and then cubic test specimens of 10 mm on each side along the coordinate axes are cut out from the region beneath the end plate by a diamond cutter (SBT650, South Bay Technology, Inc., San Clemente, CA, USA). These specimens are used for the former two tests. Single trabecula specimens are removed from the cubic specimen for the latter two tests. The detail of the choice and separation of single trabecula is described in the below. Figure 1 shows the outline of preparation of test specimens.

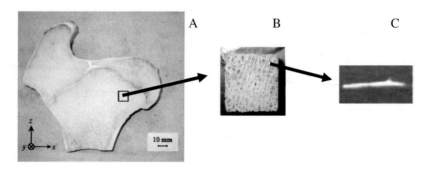

Figure 1. Preparation of specimens. A: Bovine femur sectioned by sagittal plane. B: Cubic test specimen of 10x10x10mm on each side. C: Test specimen of single trabecula removed from cubic test specimen.

2.2 Compression test

Cubic test specimens are sent to uniaxial compression test that is carried out on a light duty universal testing machine (EZTest-100N, Shimadzu, Kyoto, Japan). The test is conducted for three different directions of compression along x-, y-, and z-axes within the range of nondestructive compression load. The zero load position is identified by removing a small load of 2 N applied to the specimen. The maximum load is 40 N for the compression test at the compression speed of 3 millimeters per minute. The loading and unloading are repeated several times and the stress-strain curves are captured by a computer where the strain is calculated based on the distance between the compression plates of the testing machine. The elastic moduli, E_x, E_y and E_z as the cancellous bone are determined from the slope of stress-strain curve obtained.

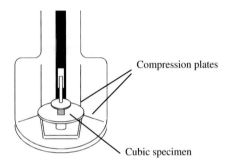

Compression plates

Cubic specimen

Figure 2. Setup for compression test of cancellous bone. The specimen is placed between the standard compression plates of a light duty universal testing machine. Three directions of compression are identical to specimen sides.

2.3 Mean intercept length analysis

Following to the compression test, cubic test specimens are sent to the three-dimensional imaging of trabecular architecture. The specimen is mounted on a micro X-ray CT-scanning device (NX-HCP-C80-I, Nittetsu Elex, Tokyo, Japan), and 150 slices images in x-y plane are scanned with the slice interval of 66 micrometers. Original slice images of 1024x1024 pixels are obtained by 600 projections with the accumulation of 32 times. As the result of the computer processing to obtain the binary images, the individual slice image is converted to 512x512 pixels of 42 square micrometers.

The mean intercept lengths [5] are calculated for each cubic specimen based on its three-dimenaional image data. The interval of the orientation of inspection line is one degree both in the longitudinal and latitudinal directions. The data of mean intercept length with orientation of inspection line are fitted by the fabric ellipsoid [6], and the principal values and direction cosines of principal axes with respect to coordinate axes are determined.

A   B

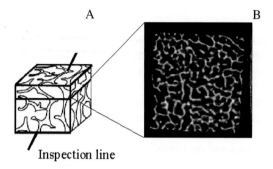

Inspection line

Figure 3. Inspection line for mean intercept length analysis. A: Inspection line for cubic specimen. B: Slice image of cubic specimen.

2.4 Bending test

By referring to the three-dimensional micro CT image of each cubic specimen, trabecula specimens are chosen for the bending test. Cubic specimen is sliced to plates of 1 millimeter thick along three different surfaces by using diamond cutter, and then trabecula specimens aligned in x-, y- and z-directions are removed from these slices by surgical blade under a microscope. Three point bending test was conduced for these specimens on the experimental setup with the support distance of 3 millimeters as shown in Figure 4. The major and minor diameters of the trabecula specimens are measured by using CCD camera, and the cross-sectional properties are calculated by assuming the elliptic cross-section. Bending was applied to the specimen three times with the maximum deflection of 40 micrometers, and the deflection and the bending force are monitored through an extensometer (EDP-5A-50, Tokyo-Sokki, Tokyo, Japan) and a load cell (LVS-100GA, Kyowa-Dengyo, Tokyo, Japan). The elastic modulus is calculated from the slope of load-deflection curve observed after second cycle based on the beam theory of three-point bending.

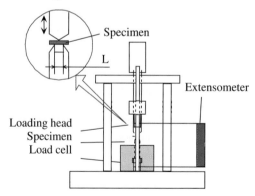

Specimen

L

Extensometer

Loading head
Specimen
Load cell

Figure 4. Set-up for three point bending of single trabecula. Beam span L is 3mm, and outputs of load cell and extensometer are connected to AD ports.

2.5 Hardness test

Another set of single trabecula specimens is prepared for Vickers hardness test. Each specimen is laid and embedded in the surface layer of plaster bed. The surface of trabecula specimen is finished by using abrasive paper and abrasive soap (Metapolish, Fujimi, Tokyo, Japan) as is shown in Figure 5A. The embedded specimen is sent to the hardness tester (HMV-1, Shimadzu, Kyoto, Japan). The shape of indentation tool is quadrangular pyramid, and the testing load of 10 grams is applied for 10 seconds to the polished surface of specimen. These conditions are determined by referring to the preliminary trials. Indentation is carried out for five points along the axis of trabecula specimen (Figure 5B), and the stamp size is measured through a CCD camera. Concerning the depth of the test surface, preliminary trials for a couple of surfaces of different depth have confirmed no difference in results of indentation test.

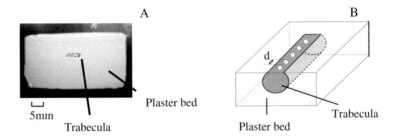

Figure 5. Trabecula specimen for hardness test. A: Finished surface of single trabecula test specimen embedded in plaster bed. B: Indentation points along londitudinal axis of tarbecula specimen.

3   Results and Discussion

3.1 Structural anisotropy

Table 1 shows the direction cosines of the principal axes of the fabric ellipsoid with respect to the coordinate axes obtained as the result of the mean intercept length analysis and fabric ellipsoid fitting. The direction cosines between the first principal axis and x-axis, the second principal axis and y-axis, and the third principal axis and z-axis are almost unity. That is, the anterior, medial/lateral and proximal directions of the femur are almost identical to the principal direction of the trabecular architecture of the region of cancellous bone tested in this study. As shown in Figure 6, the principal value of the fabric ellipsoid is the largest in the third principal direction almost identical to z-axis, and the smallest in the first principal direction almost identical to x-axis. These data show the structural anisotropy of trabecular architecture of the cubic specimen as the cancellous bone tested.

3.2   Mechanical anisotropy: cancellous bone

The anisotropic mechanical properties observed in the elastic modulus by the compression test of cubic specimen of cancellous bone are shown Figure 7.   The modulus is the smallest in x-direction, and the largest in z-direction.   As the result

Table 1.   Direction cosines of principal axes.

	$\cos\theta(x,1)$	$\cos\theta(y,2)$	$\cos\theta(z,3)$
Mean	0.961	0.960	1.000
SD	0.0425	0.0427	0.000

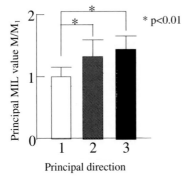

Figure 6.   Principal value of mean intercept length (MIL) representing the structural anisotropy of trabecular architecture.

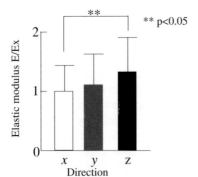

Figure 7.   Anisotropy observed in elastic compression moduli as the cancellous bone.

of the mean intercept length analysis and fabric ellipsoid fitting, the x-axis is almost identical to the first principal direction corresponding to the smallest principal value, and the z-axis to the third principal direction corresponding to the largest principal value. Though the difference of significance among the elastic moduli is moderate (p<0.05), the correspondence between the anisotropic mechanical property and the anisotropic structural property of trabrecular architecture of the cancellous bone would be an important aspect in the anisotropic characteristics of the cancellous bone.

3.3 Mechanical anisotropy: single trabecula

Figure 8 shows the anisotropy observed in the elastic modulus of single trabecula aligned in different directions in cancellous bone. The modulus observed by the bending test is the largest in the trabeculae aligned along z-axis, and is the smallest in the trabeculae aligned along x-axis. The difference between them is moderate but significant (p<0.05). The difference among the elastic moduli of trabeculae aligned along different direction is similar to the difference observed among the elastic compression moduli in different direction of the cancellous bone. That is, this similarity suggest us the influence of the material property dependent on the orientation of trabeculae on the macroscopic elastic property as the cancellous bone.

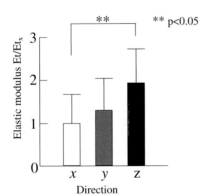

Figure 8. Anisotropy observed in elastic modulus as single trabecula aligned along different direction in cancellous bone.

3.4 Mechanical anisotropy: trabecula hardness

The dependency of Vickers hardness on the trabecula direction in the cancellous bone is shown in Figure 9, and it also exhibits the anisotropic characteristics. That is, the hardness of trabeculae aligned along z-axis is the largest and the hardness of trabeculae aligned along x-axis is the smallest. This difference is again significant (p<0.01). The dependency on the trabecula direction in the cancellous bone is

132

identical to that found in the elastic modulus of trabecula shown in the previous section. Since the positive correlation is known between the elastic modulus and the hardness of bone in macroscopic level, the result shown in this and previous sections is a reasonable finding for trabecula in microscopic level.

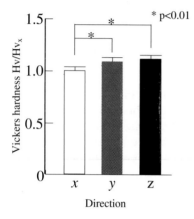

Figure 9. Anisotropy observed in Vickers hardness of single trabecula aligned along different direction in cancellous bone.

4 Summary and Remarks

This article have described a set of four mechanical tests/analysis has been described for the cancellous bone specimen of macroscopic level and the single trabecula specimen of microscopic level, and have considered the anisotropic characteristics of cancellous bone from the both levels.

Compression test and mean intercept length analysis have illustrated the mechanical and structural anisotropic characteristics of cancellous bone. Three point bending test and Vickers hardness test have illustrated the dependency of the mechanical property of trabecula on the trabecula direction in the trabecular architecture of cancellous bone. These findings, collectively, have suggested a possibility that the anisotropic mechanical property as the cancellous bone in macroscopic level is dependent on the structural anisotropy of trabecular architecture as well as the mechanical properties of trabecula in micro level dependent on the direction of trabecula in the cancellous bone in part. Further consideration will be expected to examine the anisotropy of trabecula at micro level more in detail.

Acknowledgements

This work was supported by Grant-in-Aid for Scientific Research on Priority Areas 15086210 from the Ministry of Education, Culture, Sports, Science and Technology of Japan.

References

1. Mow, V.C., Hayes, W.C., 1991, Basic Orthopaedic Biomechanics, Raven Press.
2. Cowin, S.C., 1986, Wolff's law of trabecular architecture at remodelign equilibrium, J. Biomech. Engng., 108, 83-88.
3. Gibson, L.J., 1985, The mechanical behavior of cancellous bone, J. Biomech., 18, 317-328.
4. Vahey, J.W., Lewis, J.L., Vanderby, R.Jr., 1987, Elastic moduli, yield stress and ultimate stress of cancellous bone in the canine proximal femur, J. Biomech., 20, 29-33.
5. Whitehouse, W.D., 1974, The quantative morphology of anisotropic trabecular bone, J. Biomech, 101, 153-168.
6. Harrigan, T.P., Mann, R.W., 1984, Characterization of microstructural anisotropy in orthotropic materials uisng second rank tensor, J. Mater. Sci., 19, 761-767.
7. Rho, J.Y., Liisa, K.S., Zioupos, P., 1998, Mechanical properties and the hierarchical structure of bone, Med. Eng. Phys, 20,92-102.
8. Zysset, P.K., Gou, X.E., Hoffler, C.E., Moore, K.E. Goldstein, S.A., 1999, Elastic modulus and hardness of cartical and trabecular bone lamellae measured by nanoindentation in the human femur, J. Biomech, 32, 1005-1012.

APPLICATION OF COMPUTATIONAL BIOMECHANICS TO CLINICAL CARDIOVASCULAR MEDICINE

T. YAMAGUCHI

Department of Bioengineering and Robotics, Tohoku University,
6-6-01 Aoba-yama, Sendai 980-8579, Japan
E-mail: takami@pfsl.mech.tohoku.ac.jp

Cardiovascular diseases, particularly ischemic heart disease such as myocardial infarction (MI), are the leading cause of death in the industrialized world. Vascular diseases including MI share a common background, atherosclerosis, and a common final event, the breakage or destruction of vascular structure. Both the onset and final outcome of fatal vascular diseases are related to mechanical events that occur on the vascular wall, probably owing to alterations in blood flow. Consequently, the fluid-solid mechanical interactions between blood and the vascular wall must be analyzed in order to predict, diagnose, and prevent the fatal consequences of vascular disease. We need to use computational studies to elucidate the mechanism of such disease, to refine the diagnostic measures, and to develop therapeutic modalities, either invasive or non-invasive. In this review, we discuss why computational study is necessary, how a computational model is built, the pre-requisites for computation, and the pitfalls of interpreting computational studies.

1 Introduction

Currently, there is a strong thrust to develop clinical applications of computational fluid and solid mechanics, particularly for cardiovascular medicine [1, 2]. Rapid advances in computers have made this easier. Nevertheless, there are still many problems to overcome before an actual clinical system can be constructed. Therefore, it would be useful to summarize these problems.

2 Problems in Computational Biomechanics

There are four major difficulties in the computational biomechanical analysis of the cardiovascular system: the complex geometry, complex boundaries, unsteadiness of blood flow, and blood as a non-Newtonian fluid. These have been recognized as difficult, but essential problems, since the earliest applications of the computational method. Numerous attempts have been made to solve them, although we have not yet resolved them satisfactorily. This is true not only for each category of problem; combinations of these problems complicate the matter, since each problem is essentially non-linear. Moreover, these difficulties in blood flow analysis influence each other in a non-linear way, further compounding the problem. Let us discuss some of our current ideas.

2.1 Complex geometry and modeling

Every component of the cardiovascular system, including the heart, arterial trees, and veins, has a very complex geometry. For example, the inner surface of the heart is not smooth, but is covered by trabeculae. The aorta curves and distorts markedly, changing its cross-sectional shape, especially in terms of average parameters, such as its diameter and area. Artery branches and veins merge and the branching and merging patterns are suspected of being related to the pathogenesis of vascular disorders, which are the most important causes of death in the industrialized world. Since the geometry is complex and the influence of this complex geometry cannot be reproduced exactly or evaluated using either experimental or theoretical methods, we need a computational method [3].

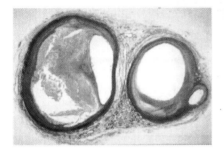

Figure 1. Distribution of atherosclerotic plaque in the carotid bifurcation. A remarkable atherosclerotic plaque occupies almost 80% of the cross-sectional area of the internal carotid artery of a human. The inner (flow divider) walls of both the internal and external artery are spared and the plaque (internal carotid) and intimal thickening (external cartotid) are seen on the outer sides of the bifurcation.

Any computational method requires models, and the complex geometry makes it necessary to model the cardiovascular system in an individual, or patient-specific, manner. Individual modeling as such can only elucidate the pathogenesis of vascular disorders through analysis from a fluid mechanics viewpoint. This is particularly true because wall shear stress is considered responsible for the physiological and pathological changes of the arterial wall (Fig. 1). The wall shear stress is extremely susceptible to minute alterations in the luminal surface of the arterial wall. Therefore, to evaluate the exact wall shear stress pattern, we need to know the geometry in great detail, and this differs from person to person.

However, we also need to be aware that the overall or global configuration of the cardiovascular system primarily determines the first order structure of the flow inside it, and hence the distribution of the physical parameters that may affect pathophysiological phenomena. In this sense, we do not know whether the instantaneous wall shear stress pattern or long-term average pattern actually governs the progress of disease. The atherosclerosis underlying cerebro- and cardiovascular events grows and worsens over very long periods, on the order of decades, in the human vascular system. This suggests that the prediction and prevention of these fatal events should be based on observations and follow-up over a very long time scale. Any computational application should incorporate this very long time scale.

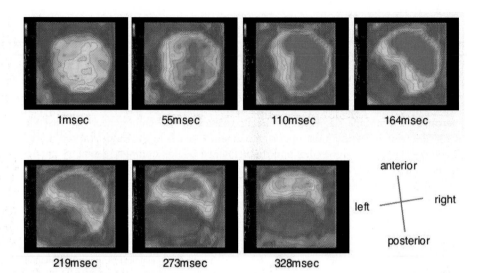

1msec   55msec   110msec   164msec

anterior

left — right

posterior

219msec   273msec   328msec

Figure 2. Measured velocity distribution profile of a healthy human aorta using the phase map MRI method (data courtesy of Prof. H. Isoada, Hamamatsu Medical University). A large-scale swirling motion of the vortex structure is seen, particularly from late systole to early diastole.

Frequently, newcomers to the field jump into direct explanations that are sometimes not correct from a biological viewpoint. The disease process may not be clearly distinguished from physiological alterations of the vascular system, so that analysis should be combined with an understanding of the normal aging process. Aging is also responsible for large-scale geometric changes in the cardiovascular system, as is the growth that forms the structures of the body in the young. Computational prediction of disease, if possible, should be based on these considerations and should consider long-term phenomena. This is true when we model the living system using modern technology, particularly imaging technology.

To date, various image-based technologies have been developed and used for the computational mechanical analysis of blood flow. They include computed tomography (CT), magnetic resonance imaging (MRI), and ultrasound methods. Of these imaging techniques, MRI methods support the greatest expectations because of their inherent non-invasive nature (Fig. 2). Non-invasiveness is very important, because healthy subjects are candidates for preventive examination in cardiovascular medicine. Methods using X-rays, such as CT, cannot be used for mass screening of a large population, although their resolution and reproducibility may be better than other methods. In any imaging technology, we obtain pixel or voxel images of the target organ, so that a method for building a computational model out of medical images has been a focus of interest for several years, and various methods have been proposed (Fig. 3) [4-6].

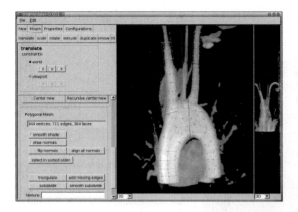

Figure 3. An example of a screen capture used for interactive modeling of the aorta and its large branches using an MRI angiographic image.

2.2 Complex boundaries

We typically analyze the blood flow field in a confined region with definite boundaries. Free boundary analysis, such as aerofoil analysis, is exceptional in blood flow analysis. Conventionally, we can distinguish at least three different kinds of regional boundary in a computational fluid dynamic analysis: inlets, outlets, and walls [7-9].

Inlet conditions are very important in engineering studies of high Reynolds number (turbulent) flow, because upstream phenomena, such as small disturbances and vortices, strongly affect the transition, and hence the nature of the whole flow field. In arterial flow, we usually assume flat or parabolic profiles for steady flow analysis and the so-called Wormersley profile or measured velocity data for unsteady analysis as inlet velocity conditions. Differences in the computed results have been examined in many cases. Our experience has confirmed that discussion of the inlet length in viscous flow can be applied to estimation of the length or depth affected by the transition. This is valid not only for inlet boundary conditions for the entire flow field but also for discussion of some wall changes. In other words, the extent of the effects of any kind of disturbing structure in the cardiovascular system is limited when the Reynolds number is relatively low. As we do not compute the blood flow using a Reynolds number greater than 1000, the viscous inlet length would be 10 times the representative length (diameter). If we look at peripheral or distal regions, the Reynolds number is about 100, and most of the inlet disturbances die out in a stream-wise direction equivalent to 1 or 2 diameters. As we have shown, the inlet portion itself is strongly affected by the upstream flow. We show a combined model of the left ventricle and aorta (Figure 4), whose velocity profiles differ from those of simple aorta models up to several diameters downstream from the aortic valve. The global configuration of the aorta, including its curvature and

138

distortion, cancel out the inlet condition after several diameter lengths along the stream.

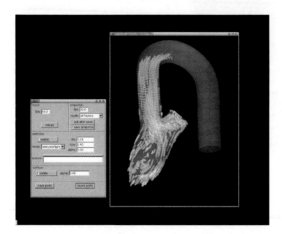

Figure 4. A particle-tracking image of the blood flow in the left ventricle and aorta using a combined model. The complex swirling motion of the fluid that forms inside the left ventricle is convected downstream to the ascending aorta. The residual swirling motion of the fluid affects the downstream flow, particularly from late systole to diastole because of the reversal of the pressure gradient that makes the entire flow field unstable (Computational results and visualizaton data courtesy of Dr M. Nakamura and Mr. T. Hayasaka).

Outlet conditions are also important, particularly when we consider the overall distribution of the blood flow in a complex branch. Although we usually assume that the arterial diameter reflects the average flow rate of the artery adaptively, this is obviously inapplicable to general physiological conditions. For example, physical exercise of muscles or activation of some organs may alter the peripheral resistance markedly, so that the distribution of the blood flow changes readily, which modifies the flow field in the upstream branching region. There have been some attempts to combine the peripheral state and outlet boundary condition by solving a one-dimensional fluid equation and combining the results for the upstream flow fields. To build a comprehensive model of the cardiovascular system that includes the entire circulation, combination of peripheral or organ-level circulation becomes an important issue. It is important to consider the venous return to the central circulatory system as well as the effect of back-propagated pressure wave reflections to the main flow field.

The third and most interesting boundary is the wall. The arterial wall is made of extremely soft material from an engineering viewpoint, although it is much stiffer than the venous wall. Elastic arteries, such as the aorta, are flexible, and their diameters alter by more than 10% with each heartbeat. Although it is believed that the global characteristics of the blood flow in the large arteries are less affected by elastic deformation of the wall, the near wall phenomenon is undoubtedly influenced by wall movement. Wall shear stress is one such near-wall phenomena and is suspected of being susceptible to wall deformation. In particular, flow separation could be affected greatly if it occurs at a site with complex geometry. However, there is no widely available, fully reliable, practical coupled analysis

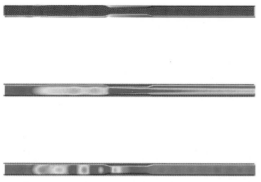

Figure 5. Wave propagation phenomena computed in a model of arterial stenosis using a fluid-solid interaction computational code. Top: wall motion, Middle: velocity distribution, Bottom: Pressure wave. Complex propagation patterns are reproduced, including splitting, reflection, acceleration and deceleration (Computational results and visualizaton data courtesy of Mr. T. Fukui.).

computational code for the study of fluid-solid interactions. This is mainly owing to the different time scales for the fluid and solid phases of the field in terms of basic mechanics, as well as technical constraints related to computational iterations. An incompressible fluid, such as water, has a much smaller time scale for the propagation of information, *i.e.*, the speed of sound is very high compared with the propagation speed of other physical phenomena, such as wall deformation, velocity fields, etc. Since we have to match the time scales of various phenomena when advancing the computational steps in the combined analysis of fluid-solid interactions, some compromise must be introduced for the faster part of the interaction. This is particularly true when we use a so-called weakly coupled method, which seems to be the only plausible approach given the current state of the art. Moreover, coupled analysis requires many additional constraints and assumptions, such as the linear material properties of the wall and neglecting external fixation (tethering) of the arterial system. This is partly why noone has obtained satisfactory computational results in this subject.

Nevertheless, it is now possible to solve fluid-solid interactions using computation, and some results for a limited number of phenomena, such as the analysis of pulse wave propagation velocity measurements, have been reported (Figure 5).

2.3   Unsteady blood flow

Since the only energy source for the cardiovascular system is the heart, which contracts and dilates during the cardiac cycle, the blood flow should be regarded as pulsatile, at least in large arteries and veins. Pulsatility introduces an intermediate-length time scale to the computational analysis. In addition, the above-mentioned characteristics, *i.e.*, the complex geometry and flexible wall, make the interactions of blood flow and wall motion more complex. Pulsatile movements and deformation of the wall pose difficult problems for computation. For example, in a complex arterial bifurcation, the pulsatility of the blood flow affects the distribution of the flow rate to downstream branches; hence, the wall shear stress distribution pattern

140

may be affected greatly. Flow separation from the wall may also be influenced, so that the reattachment and size of the vortices is affected. In addition to local fluid mechanical phenomena, the upstream condition may differ with the frequency of the pulsatility. The frequency of the major pulse, as well as the induced secondary motion of the fluid, changes the mass transfer time scale, so that the influence of the flow on the physiology and pathology of the wall may be altered completely.

Note that the flow field evolves in the computation, so that several heartbeats should be included to obtain stable results. As with the geometrical consideration and boundary conditions, the frequency of the real heartbeat varies very markedly. Since Wormersley's alpha parameter is proportional to the square root of the frequency, the more the frequency increases, the less viscous the flow field becomes. Therefore, tachycardia enhances the influence of the global configuration of the cardiovascular system. The effect of the heartbeat has not been investigated fully, particularly in the context of computational fluid dynamics, and could be of interest when discussing the preventive evaluation of various physiological conditions after developing a practical analysis system.

In addition, little attention has been paid to much longer time scales, including growth and remodeling in the course of growth, and the even longer time scale of the evolutionary process. In the latter case, comparative biological viewpoints should also be considered carefully and should help in understanding the computational results. In this context, the configuration of the cardiovascular system is a complex combination of the results of physiological and pathological processes over an extremely wide range of time scales, from evolutionary to quantum-level fluctuations.

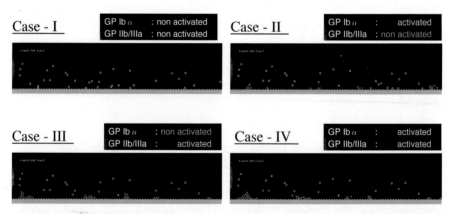

Figure 6. Simulation of platelet adhesion (primary thrombus formation) in the blood flow near the vascular wall. Particle motion was simulated using a discrete element method and fluid motion was simulated using a Stokesian dynamics method. Simulating activation of receptors that govern adhesion, the mechanisms of the formation, and destruction of a primary thrombus were analyzed.

2.4 Blood as a non-Newtonian fluid

It is frequently asked whether the non-Newtonian viscosity affects the flow field in arteries and veins. Of course, it affects the local and global nature of the blood flow to some extent, although the influence has not been fully appreciated owing to computational restrictions. Roughly speaking, there are two main possible approaches for dealing with the non-Newtonian nature of blood. One is based on the conventional continuum approach, which assumes a constitutive relationship, such as the exponential viscosity law. Although this approach considers many theoretical and experimental problems, we do not discuss it further here. The other considers the blood to be composed of multi-phase substances and introduces various levels of sub-modeling of the behavior of the blood components. This method is currently studied actively.

In this approach, we can introduce subclasses depending on the manner and level of modeling. The recent application of so-called particle methods, such as smoothed particle hydrodynamics (SPH) and a moving particle semi-implicit (MPS) method to the blood flow is one subclass. In this method, continuous fluid is expressed using particles that interact physically. This is also a kind of mesh-less method to which constitutive laws can be introduced in terms of collision and repulsion among particles. This is a promising method because of its intrinsic Lagrangean approach, which makes it easy to analyze fluid-solid interactions.

Another way of thinking is possible because the blood is composed of particles and fluid [10]. For example, platelets have the ability to form thrombi under the influence of the blood flow. Red blood cells can deform markedly as they flow in capillaries together with white blood cells. Such biological responses can be modeled properly using models of physical and even chemical interactions between the particles and between each particle and other substances and structures. Compared with the simple particle method, we can call the latter method the biological particle method, which we believe can open a totally different computational scheme that can describe and model non-Newtonian physical laws, as well as biological constitutive laws, to promote further computational mechanical analysis of blood flow (Figure 6).

3 Concluding Remarks

In concluding this article on the computational mechanics' approach to cardiovascular fluid-solid interaction studies for developing clinical applications, we would like to point out the necessity of being aware that many theoretical and fundamental questions remain unsolved. With advances in computing technology, massive computing power has become available. This definitely makes it easy to conduct extremely large-scale computing. However, we must always remember that the system that we are to solve is truly biological and the biological system is alive.

142

What we hope to accomplish is necessarily biology, even if it appears to be a purely mathematical or physical problem. A living system always responds to its environment, and can remodel itself in order to accommodate a necessary reactive mechanism, and evolve through reproduction over a longer time scale. Although mechanical (*i.e.*, numerical) accuracy is undoubtedly important, the biological considerations remain paramount.

Acknowledgements

This work was supported by Grant-in-Aid for Scientific Research on Priority Areas 15086204 from the Ministry of Education, Culture, Sports, Science and Technology of Japan.

References

1. Yamaguchi, T., 2000. Clinical Application of Computational Mechanics to the Cardiovascular System, Springer-Verlag, Tokyo, Berlin, Heidelberg, New York.
2. Yamaguchi, T., Hayasaka, T., Mori, D., Hayashi, H., Yano, K., Mizuno, F., Harazawa, M., 2003. Towards computational biomechanics-based cardiovascular medical practice. In: Armfeld, S.W., Morgan, P., Srinivas, K. (Eds.), Computational Fluid Dynamics 2002, Springer-Verlag, Berlin, Heidelberg, New York, pp. 46-61.
3. Yamaguchi, T., Taylor, T.W., 1994. Biomechanics of atherosclerosis and the blood flow. In: Hirasawa, Y., Woo, S.L-Y., Sledge, C.B. (Eds.), Clinical Biomechanics. Springer-Verlag, Tokyo, pp. 225-238.
4. Hayashi, H., Yamaguchi, T., 2002. A simple comutational model of the right coronary artery on the beating heart-effects of the temporal change of curvature and torsion on the blood flow. Biorheology 39, 395-399.
5. Mori, D., Yamaguchi, T., 2002. Computational fluid dynamics modeling and analysis of the effect of 3-D distortion of the human aortic arch. Computer Methods in Biomechanics and Biomedical Engineering 5, 249-260.
6. Mori, D., Hayasaka, T., Yamaguchi, T., 2002. Modeling of the human aoritic arch with its major branches for computational fluid dynamics simulation of the blood flow. JSME International Journal Series C 45, 997-1002.
7. Yamaguchi, T., Taylor, T.W., 1994. Some moving boundary problems in computational bio-fluid mechanics . In: Crolet, J.M., Ohayon, R. (Eds.), Computational Methods for Fluid-Structure Interaction. Longman Scientific & Technical, Harlow, U.K., pp. 198-213.
8. Yamaguchi, T., 1996. Computational visualization of blood flow in the cardiovascular system. In: Jaffrin, M.Y., Caro, C.G. (Eds.), Biological Flows. Plenum Press, New York, pp. 115-136.

9. Yamaguchi, T., 1996. Computational visualization of blood flow, In: Hayashi, K., Ishikawa, H. (Eds.), Computational Biomechanics. Springer-Verlag, Tokyo, pp. 165-184.
10. Miyazaki, H., Yamaguchi, T., 2002. Formation and destruction of primary thrombi under the influence of blood flow and von Willebrand factor analyzed by a discrete element method. Biorhelogy 40, 265-272.

BIOMECHANICAL STUDY FOR SKELETAL MUSCLE INJURY AND A VIEW OF MICRO-BIOMECHANICS FOR MICROSTRUCTURE OF MUSCLE

S. YAMAMOTO AND E. TANAKA

Department of Mechanical Science and Engineering, Nagoya University, Furo-cho, Chikusa-ku, Nagoya 464-8603, Japan
E-mail: sota@mech.nagoya-u.ac.jp

This paper concerns with the prediction and prevention of skeletal muscle injury. To clarify the mechanism of muscle injury, we conducted biomechanical and pathological evaluations for muscle contusion and strain-injury. The results showed correlations between severity of pathological damage and functional disability on contraction. The results also suggested that microscopic damage of muscle fibers, peripheral circulation or motor units has significant effects on the change of macroscopic function, and that microscopic examination is prerequisite to understand these phenomena. As a preliminary study, previous papers on microstructural elements of muscle are reviewed.

1 Introduction

Skeletal muscle injuries are frequently observed in traffic and sports accidents and classified into several injury types depending on the cause of injury. Muscle contusion is caused by impact compression normal to the muscle fiber while strain injury is caused by sudden stretch. It is important to clarify the mechanism of such injuries by considering microstructure of muscle tissue for injury prevention and accident reconstruction. However macroscopic correlations between mechanical loading and muscle injury have not been discussed enough in the current situation.

In this paper, we performed macroscopic impact tests for rabbit tibialis anterior muscle to simulate muscle contusion or strain-injury. Pathological and mechanical evaluations for such muscle injuries were conducted. Then we reviewed studies on microstructure of skeletal muscle and discussed a view of micro-biomechanics study for muscle injury prevention. These works will help us to facilitate the elucidation of micromechanism of muscle injury.

2 Methods

We used tibialis anterior (TA) muscle of female Japanese white rabbit. Twelve rabbits (2.92 ± 0.12 kg, mean $\pm$ S.D.) were used for muscle contusion tests, 21 (2.94 ± 0.17 kg) were for strain injury tests and other 7 (2.88 ± 0.20 kg) were used as the control.

2.1 Muscle contusion test

Firstly, we performed muscle contusion tests. Crisco et al. performed impact tests using gastrocnemius muscle but they impacted muscles percutaneously [1]. So it was not easy to evaluate the impact energy applied to the muscle quantitatively. Therefore we impacted muscle directly using the apparatus shown in Figure 1.

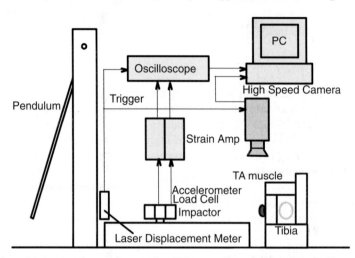

Figure 1. Impact test system for muscle contusion. This system has an impactor launched by a pendulum. The impacter has a load cell (LUR-A-2KNSA1, Kyowa) to measure the impact force and the data were recorded by a digital oscilloscope (TDS420A, Sony Tektronix). The impact velocity was calculated from a movie recorded by a digital high-speed camera (MEMRECAM fxK3, NAC Image Technology).

The following procedure was used for the preparation. The animals were anesthetized by pentobarbital sodium (30 mg/kg) or isoflurane, after which the TA muscle and deep peroneal nerve were exposed. Two nigrosin markers were added on the surface of the proximal end of the muscle belly and distal tendon to determine the muscle length. The *in situ* length was defined as the distance between two markers at a 90-degree angle to the ankle joint. Next we drilled into the tibial condyle, and a Kirschner's wire (ϕ = 7 mm) was inserted into the hole. Finally, the distal tendon of the TA muscle was cut. The wire in the tibial condyle was fixed on the holder, and the distal tendon was directly gripped by a tooth-like jig.

The muscle was cyclically stretched with 7 mm amplitude (less than 13% strain) at the velocity of 7 mm/sec as a preconditioning. Next we performed an isometric contraction test to evaluate the initial contraction force of the specimen. The specimen was activated by an electric pulse with a frequency of 50 Hz during 10 seconds. The voltage of the pulse required for maximum contraction was determined as tenfold the threshold for twitching. A thin wire electrode for EMG (ST. Steel 7 Strand, Teflon, A-M Systems) was inserted into the distal end of the muscle belly, and another electrode was directly clipped onto the deep peroneal

146

nerve. Then the muscle was impacted with an impactor (250 g weight, 5 mm width of impact surface).

For a pathological evaluation, the muscle was sliced and stained with hematoxylin and eosin after soaking in formalin more than 1 week. Then we examined severity of muscle contusion microscopically. As mechanical evaluation of muscle contusion, we conducted isometric and tensile tests of injured muscles by the apparatus shown in Figure 2. The injured muscle was elongated at 200 mm/sec tensile velocity until break in the activated condition.

2.2 Strain injury test

We also performed strain injury tests using the apparatus shown in Figure 2. Since a linear motor type actuator performs with high acceleration and accurate position control, this type of actuator is appropriate to make an impact stretch to induce strain injury. The procedure of the strain injury test was similar with that of the contusion test except for the induction process of injury. As the induction process of injury we applied loading/unloading process of stretching at 200 mm/sec to the specimen with maximally activated condition. Applied stretch was 20, 25 or 30 % of the in situ length based on the results of Hasselman et al. [2]. Pathological and mechanical evaluations were done as same procedure as the muscle contusion test.

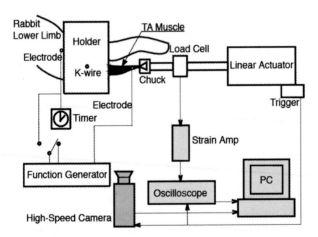

Figure 2. Strain injury test system. This apparatus consists of a linear motor type actuator (GLM-20, THK), a knee holder, a load cell (LUR-A-2KNSA1, Kyowa) a digital high-speed camera (MEMRECAM fxK3, NAC Image Technology), and a digital oscilloscope (TDS420A, Sony Tektronix). The load cell detects the load on the muscle and the data were acquisitioned on the digital oscilloscope. The elongation of the muscle was calculated from the distance between the nigrosin markers on the surface of the muscle recorded by the digital high-speed camera.

3 Results and Discussion

3.1 Muscle contusion

Results of the pathological evaluations for the muscle contusion tests were shown in Figure 3. The severity of injury could be divided into 3 grades named as Grade I, II and III. Grade I is the case in which no blood bleeding and failure of muscle fiber are observed. Grade II is the case in which sporadic bleeding is observed. In Grade III we observe severe damage spread to whole tissue. Based on these injury grades, the relation between the impact energy and injury severity was evaluated as shown in Figure 4. We observe that the two thresholds may exist between Grade I and II and between Grade II and III.

Figure 5 shows the relation between the changes of the isometric contraction force and the impact energy. The isometric contraction force dropped suddenly at the impact energy of about 0.13 J. According to Figure 4, this sudden drop is induced in the region of Grade II. The results of the tensile tests, furthermore, showed that the muscles broke at the muscle belly at the energy above 0.13 J, while the muscles applied at less than 0.13 J broke at the muscle-tendon junction (MTJ).

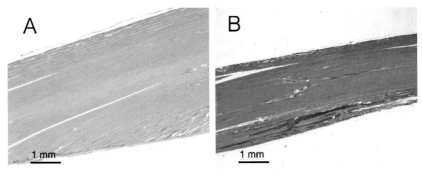

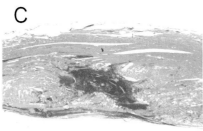

Figure 3. Classification of tissue damage for muscle contusion. A: Cases in which no blood bleeding and failure of muscle fiber are observed. (Grade I). B: Cases having sporadic bleeding. (Grade II). C: Cases having severe damage spread to whole tissue. (Grade III).

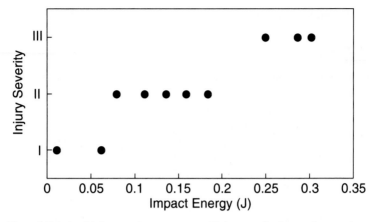

Figure 4. Relationship between impact energy and injury severity for muscle contusion.

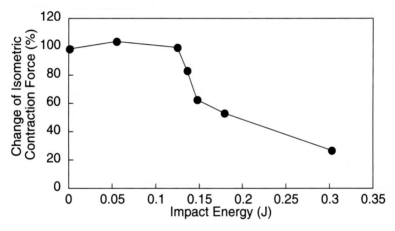

Figure 5. Relationship between impact energy and change of isometric contraction force. The threshold of impact energy for decrease of isometric contraction force can be observed around 0.13 J.

Figure 6 shows the relationships between the impact energy and the site of rupture in the tensile tests, change of isometric contraction force and the injury severity. Damaged muscle had a tendency to break at the muscle belly and the threshold of impact energy was the same level of change of contraction force.

These evaluations were conducted in five minutes after impact. Therefore the results are considered to show the instantaneous response of muscle contusion. In our microscopic observations, the ratio of the damaged fascicles was a few percent of the number of all fascicles in Grade II injury. According to Figure 5, the decrease of contraction force of the muscle with Grade II injury was much larger than the decrease expected from the results of the pathological evaluations. It indicates that macroscopic mechanical response is governed by not only the ratio of damaged muscle fiber but also the damage of perimysium or sarcolemma, which break the continuity of tissue to transmit the contraction force, and the damage of motor neuron. Since the classification shown in Figure 4 was mainly based on the muscle fiber breaking, the injury severity could not correspond with the threshold of the change of the contraction function.

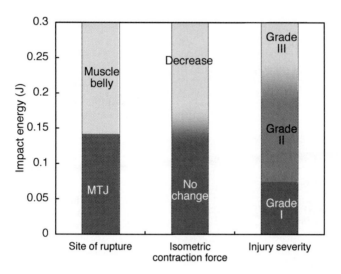

Figure 6. Relationships between the impact energy and the site of rupture in the tensile tests, change of isometric contraction force and the injury severity.

3.2   Strain injury

Next we discuss the results of muscle strain injury tests. Results of the pathological evaluations for the muscle strain injury tests were shown in Figure 7. Strain injury severity was also divided into 3 grades named as Grade 1, 2 and 3. Grade 1 means no injury. Grade 2 is the case in which some small inter-fiber blood bleeding is observed. In case of Grade III, blood bleeding and failure of muscle fibers spread over whole tissue.

The relationship between the impact energy and injury severity was shown in Figure 8. In the range between 23 and 25% stretch, we observed both Grade 2 and 3. As shown in Figure 9, when we examined the relationship between the injury severity and the absorbed energy, which is defined as the dissipation of energy during the induction process of injury, we can say that the pathological threshold of strain injury exists around 0.25 J, and Grade 2 injury is a transient state between

Grade 1 and 3. The isometric contraction force tended to decrease with the increase of the absorbed energy as shown in Figure 10. The threshold of decrease of

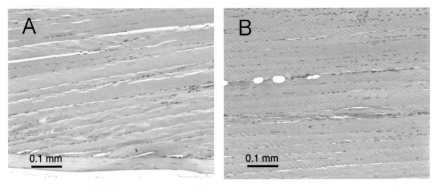

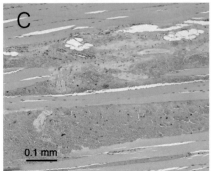

Figure 7. Classification of tissue damage for strain injury. A: Cases in which no blood bleeding and failure of muscle fiber are observed. (Grade 1). B: Cases which have some small inter-fiber blood bleedings. (Grade 2). C: Cases which have blood bleedings and failures of muscle fibers spread over whole tissue. (Grade 3).

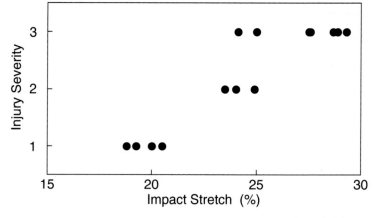

Figure 8. Relationship between impact stretch and injury severity for strain injury.

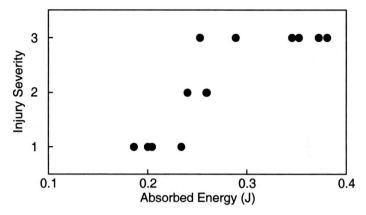

Figure 9. Relationship between the applied energy during the induction process of injury and injury severity for strain injury.

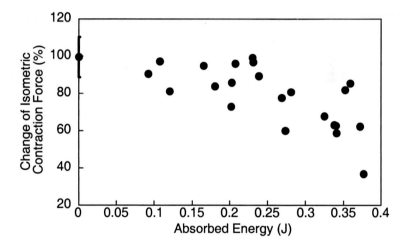

Figure 10. Relationship between absorbed energy during the induction process of injury and change of isometric contraction force.

contraction force also existed around 0.25 J. We will contine to discuss the correlation between the pathological and the biomechanical definitions of strain injury.

The results of tensile tests for damaged muscle, on the other hand, did not show any significant differences against the control group in their failure load and elongation. Therefore strain injury does not cause any significant damage on the mechanical properties of the passive structural elements, because the failure properties of muscle mainly reflects the passive mechanical properties of muscle.

4   View of Micro Biomechanics of Muscle for Injury Prevention

A skeletal muscle has a hierarchical structure that consists of sub-scale fibrous structures such as muscle fascicles or muscle fibers. A muscle fascicle is a bundle of dozens of muscle fibers surrounded by perimysium. A muscle fiber is a myocyte, which is the fundamental system of active contraction and composes of a motor unit combining with a motor neuron.

Thus many studies on muscle fibers have been conducted and experimetal apparatus and technique have been developed. Gordon et al. [3], for example, developed an experimental apparatus for isometric and isotonic contraction tests of muscle fiber. Later Ford et al. [4] developed a technique to fix a muscle fiber on a test system using Al foil, which has been used by other researchers. With such techniques, physiological response of muscle fiber such as contraction against stretch have been studied in detail (for example Mutungi et al. [5]).

A few studies, on the other hand, have been reported on injury or damage of muscle microstructure and their influences on mechanical properties or functions. Macpherson et al. [6] studied effects of injury on the contractile function of sarcomere using permeabilized muscle fiber. They concluded that the longest one among a series of sarcomeres could be damaged when the sarcomere length in the fiber is heterogeneous. Their results are quite suggestive to clarify the mechanism of muscle injury, however the injury mechanism could be influenced by damage of sarcolemma and other microstructures, such as motor neurons and peripheral circulation. Thus we need further examination for mechanical properties of other microstuctural elements.

We also need to discuss a new mechanical model of muscle for injury simulation by considering its microstructure. Yucesoy et al. [7] suggested an interesting model, which describes a muscle as two-domain finite element model. Each finite element represents a segement of a bundle of muscle fibers, surrounding connective tissues and the links between them. Using this model, they showed the importance of the myofascial force transmission pathways. Such a model that takes into account mechanical properties of two or more microstructural components can be extended to describe damage of microstrucutral elements and degradation of mechanical properties and functions of muscle tissue.

5   Conclusions

Direct compressive impact and axial stretch for skeletal muscle were conducted. Microscopic examinations of muscle contusion and strain injury were done to discuss the relation between severity of muscle injury and mechanical impact. From the results the relation between pathological change and mechanical change of muscle tissue caused by mechanical impact was obtained. A view on micro biomechanics of skeletal muscle was also discussed for applying to injury prevention.

Acknowledgements

We thank Drs. Shogo Tokudome and Masahito Hitosugi for their great contributions and advices for pathological evaluations, and Biomechanics laboratory of Toyota Central R&D Labs., Inc. for their collaborations for the development of the experimental apparatus, and Mr. Atsushi Hayakawa, Tsuyoshi Taniguchi and Daisuke Ito for their great efforts for animal experiments. This work was supported by Grant-in-Aid for Scientific Research on Priority Areas 15086208 from the Ministry of Education, Culture, Sports, Science and Technology of Japan.

References

1. Crisco, J.J., Jokl, P., Heinen, G.t., Connell, M.D., Panjabi, M.M., 1994. A muscle contusion injury model. Biomechanics, physiology, and histology, Am. J. Sports Med. 22, 702-710.
2. Hasselman, C.T., Best, T.M., Seaber, A.V., Garrett, Jr., W.E., 1995. A threshold and contimuum of injury dering active stretch of rabbit skeletal muscle, Am. J. Sports Med. 23, 65-73.
3. Gordon A.M., Huxley, A.F., Julian, F.J., 1966. Tension development in highly stretched vertebrate muscle fiberes, J. Physiol. 184, 143-169.
4. Ford, L.E., Huxley, A.F., Simmons, R.M., 1977. Tension responses to sudden length change in stimulated frog muscle fiberes near slack length, J. Physiol. 269, 441-515.
5. Mutungi, G., Ranatunga, K.W., 2001. The effects of ramp stretches on active contractions in intact mammalian fast and slow muscle fibres, J. Muscle Res. and Cell Motiolity 22, 175-184.
6. Macpherson, P.C.D., Dennis, R.G., Faulkner, J.A., Sarcomerer dynamics and contraction-induced injury to maximally activated single muscle fibres from soleus muscles of rats, J. Physiol. 500, 523-533.
7. Yucesoy, C.A., Koopman, G.H.F.J.M., Huijing, P.A., Grootenboer, H.J., 2002. Three-dimensional finite element modeling of skeletal muscle using a two-domain approach: linked fiber-matrix mesh model, J. Biomech. 35, 1253-1262.

MECHANICAL BEHAVIOR AND STRUCTURAL CHANGES OF CELLS SUBJECTED TO MECHANICAL STIMULI: DEFORMATION, FREEZING, AND SHOCK WAVES

H. YAMADA, H. ISHIGURO AND M. TAMAGAWA

Department of Biological Functions and Engineering, Graduate School of Life Science and Systems Engineering, Kyushu Institute of Technology, 2-4 Hibikino, Wakamatsu-ku, Kitakyushu 808-0196, Japan
E-mail: yamada@life.kyutech.ac.jp

The responses of cells to three mechanical stimuli were examined. (1) Two types of vascular endothelial cell finite element models were created and validated under stretched substrate conditions. The numerical simulations achieved good reproducibility with the solid model and demonstrated a potential for further development with the fluid-filled model. (2) The histological change in the tissues caused by freezing and thawing was investigated to clarify the influence of bioheat transfer parameters on tissue microstructures. (3) Cell damage due to plane shock waves was investigated experimentally, and the frequency responses of one and two cells in water were evaluated by constructing a mathematical model. The results showed that the structural effects due to the deformation processes were quite large at particular frequencies

1 Introduction

The characteristics of cells are not determined solely by DNA or genetic material. Intracellular responses occur in mechanical, electrical, and chemical fields. Therefore, it is quite important to investigate cellular responses from a variety of mechanical viewpoints, including solid and fluid mechanics, as well as heat and mass transfer.

The description of stress/strain fields in cells is one of the most important cellular issues in solid mechanics [1-3]. These states influence the signal pathways and morphology of the nucleus and cytoplasm. In recent years, hyperelastic models of cytoplasm and a nucleus have been developed and validated. These models reproduce the cytoplasm strain states and nucleus deformation well [1, 2].

Determining adequate conditions for freezing and thawing in cells is another important issue. Freezing is a typical stimulus in heat and mass transfer. It has two contrary effects on biological cells and tissues: protection and destruction [4, 5]. The former effect is used for cryopreservation, and the protection of cells, tissues, and foods. The latter effect is using in cryosurgery and the destruction of tumors. Investigating changes in the histological microstructures and mechanical properties of tissues through freezing, and the relationship between the two changes is interesting from both fundamental research and application viewpoints.

The investigation of damage mechanisms in cells is also important. Extracorporeal shock wave lithotripsy (ESWL) has become pervasive in medicine [6]. Some engineering studies have examined the effects of shock waves on living tissue cells. For example, Teshima *et al.* [7] investigated the relationship between DNA damage and maximum applied pressure. Despite many medical and engineering studies, the cell damage mechanisms due to shock waves have not been identified.

In this study, we compare two types of cell structure finite element models to elucidate the role and mechanical properties of cell components, such as the cell membrane and cytoplasm. The histological tissue change with freezing and thawing is investigated to clarify the influence of bioheat transfer parameters on tissue microstructures. The damage caused by plane shock waves to living cells (red blood cells and cancer cells) is also described. Finally, the frequency response of an elastic shell model for both single and two interacting cells in water is evaluated by constructing a mathematical model.

2 Modeling and Numerical Simulation of the Mechanical Behavior of a Vascular Endothelial Cell under Substrate Deformation

2.1 Method

Two models were compared to evaluate their descriptions of mechanical behavior in a vascular endothelia cell. The first was a hyperelastic solid model, in which the nucleus and cytoplasm were modeled as hyperelastic materials (see Fig. 1(a)) [2]. The second was a fluid-filled hyperelastic membrane model that consisted of a hyperelastic membrane surrounding a fluid-filled cavity (see Fig. 1(b)) [3]. In both models, the hyperelastic part was modeled as a neo-Hookean material that had the same stress-strain relationship as an isotropic linear elastic material in the infinite strain range.

The bottom surface of both cell models adhered to the substrate surface. The mean values of Young's modulus determined by Caille *et al.* [1], *i.e.*, 5100 and 775 Pa, were chosen for the nucleus and cytoplasm, respectively. The material constant of the cell membrane was the same as that of the cytoplasm, and the material constant of the substrate was 775 kPa. The bulk modulus of the fluid was 2 GPa, which represents the compressibility of water. Then, pure uniaxial stretching in the X direction with 4% strain (deformation of the substrate in the Y direction was constrained) was applied to the substrate to check the deformation, strain, and stress predicted by the cell models.

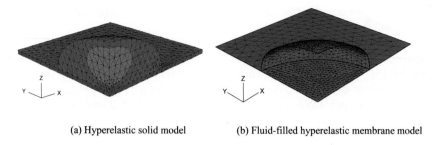

(a) Hyperelastic solid model   (b) Fluid-filled hyperelastic membrane model

Figure 1. Finite element models.

2.2   Results and discussion

Figure 2 compares the outer surface geometry predicted by the cell models at the XY and YZ symmetric planes. The fluid-filled model cell showed concavity and increased height due to the stretched substrate. These behaviors were not observed in the hyperelastic solid model. The change in cell height has not been measured because of the difficulty maintaining a suitable level of accuracy.

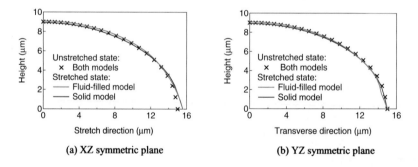

(a) XZ symmetric plane   (b) YZ symmetric plane

Figure 2. Cellular shapes for the two models shown in Fig. 1 before and after deformation under 4% pure uniaxial stretching.

Figure 3 compares the maximal principal strain near the outer surface of the predicted cellular membranes under 4% pure uniaxial substrate stretching. The strain distribution in the solid model had a directional dependence, while this effect was not as pronounced in the fluid-filled model because the solid transmitted the substrate deformation, but not the fluid or membrane.

Figure 4 compares the maximum principal stress predicted by the cell models near the outer surface under 4% pure uniaxial substrate stretching. The stress distributions were similar to the strain distributions shown in Fig. 3.

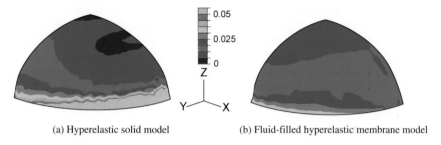

(a) Hyperelastic solid model   (b) Fluid-filled hyperelastic membrane model

Figure 3. Comparison of the maximum principal strain distributions in the cellular membrane predicted by the two models shown in Fig. 1 under 4% pure uniaxial stretching.

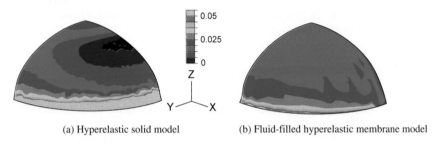

(a) Hyperelastic solid model   (b) Fluid-filled hyperelastic membrane model

Figure 4. Comparison of the maximum principal stress distributions in the cellular membrane predicted by the two models shown in Fig. 1 under 4% pure uniaxial stretching.

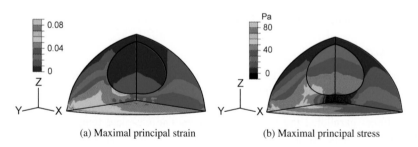

(a) Maximal principal strain   (b) Maximal principal stress

Figure 5. Maximum principal strain and stress distributions predicted by the hyperelastic solid model under 4% pure uniaxial stretching.

Figure 5 shows the maximum principal strain and stress distributions predicted by the hyperelastic model. These distributions indicated that the strain/stress decreased with increasing cytoplasm height, and the strain in the nucleus was less than that in the cytoplasm at the same height. However, the stress in the nucleus exceeded that in the cytoplasm at the same height due to the high Young's modulus. The volumetric strain and static fluid pressure of the fluid-filled cytoplasm in the fluid-filled hyperelastic membrane model were 2.3 pico-strain and 5 mPa, respectively, both of which were small enough values to provide for no effective strain/stress in the cytoplasm. Such different responses between the two types of models were caused by the lack of a cytoskeletal structure in the fluid-filled model.

3   Morphological Change of the Tissue Histological Microstructures due to Freezing and Thawing

This chapter describes the experimental materials, apparatus, methods, and results of a study of the histological change in tissues with freezing and thawing using different thermal protocols.

3.1   Experimental materials

The tissue studied was the second pectoral muscle (fresh tender meat) of chickens. This consists of muscle fibers connected by connective tissue. One muscle fiber and one cell consisted of myofibrils. The tissues were taken from a broiler (raised by Arbor Acres) immediately after slaughter and stored at 4°C for about one hour. Then, a $10 \times 10 \times 4$ mm sample was removed carefully using a microtome blade. This sample was steeped in physiological saline.

3.2   Experimental apparatus and methods

3.2.1   Cryostage

The tissues were frozen and thawed on a cryostage at a uniform temperature. The stage consisted of an aluminum block and a 1.5-mm-thick copper plate. The sample was set between a glass microslide and a glass coverslip so that the muscle fibers paralleled the cryostage surface. The Cu-plate temperature was controlled to change from room temperature to –50.0 °C at a predetermined cooling rate H and warming rate W.

3.2.2   Thermal protocols during freezing and thawing

Four different thermal protocols were followed during freezing and thawing: a) slow cooling ($H = 1.0$ °C/min) and rapid warming ($W \sim 100$ °C/min), b) rapid cooling ($H \sim 100$ °C/min) and rapid warming ($W \sim 100$ °C/min), c) slow cooling ($H = 1.0$ °C/min) and slow warming ($W = 1.0$ °C/min), and d) rapid cooling ($H \sim 100$ °C/min) and slow warming ($W = 1.0$ °C/min). The sample was maintained at the minimum temperature for 15 minutes.

3.2.3 Histological investigation of tissues

During freezing and thawing, after formalin fixation and paraffin embedding, the tissues were investigated histologically using hematoxylin and eosin (HE) staining. A slice of the sample was observed microscopically to compare the histological changes in the tissues between the thermal protocols.

3.3 Results and discussion

Figure 6 shows representative results for HE-stained tissues for each thermal protocol. In HE staining, the cytoplasm stains pink, connective tissues stain deep pink, and nuclei stain deep violet.

Before freezing (Fig. 6a), the muscle fibers (control) were compact and well-ordered. By contrast, different thermal protocols caused different changes in the histological tissue microstructures [8, 9].

After slow cooling and rapid warming (Fig. 6b), the muscle fibers had large cracks between them, and the connective tissues were separated from the muscle fibers. The muscle fibers shrank with some deformation. The effects of extracellular freezing were evident.

After rapid cooling and rapid warming (Fig. 6c), numerous small, fine cracks were formed in the muscle fibers. These were caused by intracellular ice crystals (intracellular freezing). The effect of extracellular freezing was also evident from the cracks between the muscle fibers, although these cracks were narrower than those observed after slow freezing and rapid warming. The difference in the histologically changes seen in Figs. 6b and 6c resulted from freezing at different cooling rates.

The change in the slow cooling and slow warming sample (Fig. 6d) was similar to that of the slow cooling and rapid warming sample. The muscle fibers had large cracks between them, and connective tissues were separated from the muscle fibers. However, deformation and meandering were noticeable in the fibers. In addition, the muscle fibers had a few openings between them, due to large intercellular ice crystals and cracks. Of the four cases, the most pronounced histological change in the muscle tissues was with this thermal protocol.

The change in the rapid cooling and slow warming sample (Fig. 6e) was similar to that of the rapid cooling and rapid warming sample: numerous small, fine cracks appeared in the muscle fibers, caused by intracellular freezing. However, the openings were somewhat larger than observed in the rapid cooling and rapid warming sample (Fig. 6c). This difference was caused by the different warming rate. During slow warming following the rapid cooling, the fine ice crystals in the muscle fibers became coarser during the recrystallization process. Moreover, the deformation of the muscle fibers was somewhat greater than that observed for the rapid cooling and rapid warming sample.

160

3.4   Summary

During freezing and thawing, the tissues showed remarkable changes in their histological microstructure, depending on the thermal protocols. The characteristic change for slow cooling was the formation of large cracks between the muscle fibers, caused by extracellular freezing. The characteristic change for rapid cooling was the formation of numerous small, fine cracks in the muscle fiber, caused by intracellular freezing. These two types of histological change damaged the tissues. Slow warming tended to promote the deformation and meandering of muscle fibers. These changes in the histological microstructure likely affect the mechanical properties of the tissues. This should be investigated using a biosolid mechanics approach.

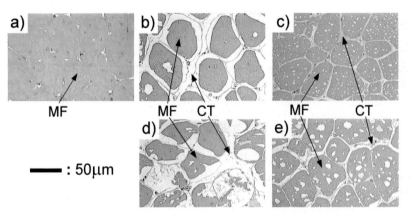

Figure 6. Tissues stained using hematoxylin and eosin after freezing and thawing (vertical sections of the muscle fibers). MF and CT denote muscle fibers and connective tissue, respectively. a) Control (before freezing), b) slow-cooling and rapid-warming ($H = 1.0$ °C/min, $W \sim 140$ °C/min), c) rapid-cooling and rapid-warming ($H \sim 90$ °C/min, $W \sim 130$ °C/min), d) slow-cooling and slow-warming ($H = 1.0$ °C/min, $W = 1.0$ °C/min), and e) rapid-cooling and slow-warming ($H \sim 90$ °C/min, $W = 1.0$ °C/min).

4   Response of Suspended Cells to Shock Waves

4.1   Cell damage caused by shock waves

Using an experimental apparatus for shock-induced damage tests (a free piston shock tube), red blood cells and general animal cells were damaged using shock waves [10]. Figure 7 shows the degree of damage to red blood cells and animal cells with a shock wave at the maximum pressure applied (surface cells from a human blood vessel). Figure 7(a) indicates that the higher the maximum pressure, the greater the degree of damage. Comparing Figs. 7(a) and 7(b), the degree of damage to the red blood cells was much less than damage to the animal cells.

One of the reasons for the large difference was the difference in the internal cell structure of the red blood cell and animal cell. Animal cells contain nuclear and intermediate filaments, while red blood cells lack nuclear filaments. Their structure depends primarily on the properties of the intermediate filament, which is made of proteins.

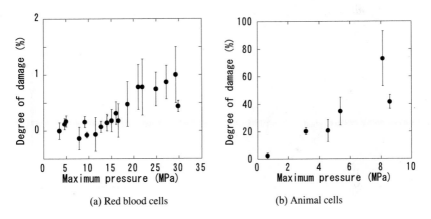

(a) Red blood cells   (b) Animal cells

Figure 7. Degree of damage to red blood cells and animal cells caused by to shock waves at maximum pressure.

4.2   Shock wave deformation model for two cells with internal structures

To investigate the effects of the internal cell structure, the deformation of red blood cells and animal cells was modeled mathematically. Suppose that one or two spherical elastic shells filled with liquid are suspended in water, and that the plane shock wave propagates from externally (Fig. 8). The radial and tangential displacements and pressure are linked for these oscillation phenomena due to a shock wave. The coupling equations for the deformation oscillation and pressure in the normal and tangential directions for each cell are

$$\frac{d^2 w_m}{dT^2} + A_{1m} w_m - A_{2m} u_m = -M \left(\Pi_m \left(1, R, T\right) - \Pi_{int} \right) \tag{1}$$

$$\frac{d^2 u_m}{dT^2} + A_{1m} u_m - A_{2m} w_m = 0 \tag{2}$$

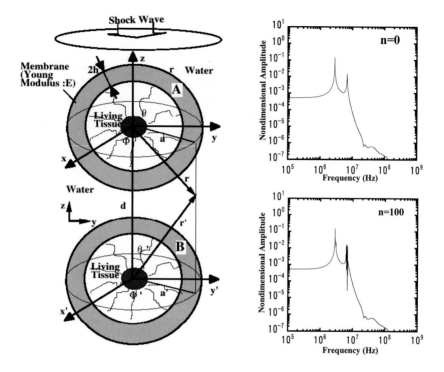

Figure 8. Mathematical model for deformation process analysis.

Figure 9. Effects of the nonlinear parameter n on the frequency response of two cells.

where w_m, u_m, Π_m, Π_{int}, and M are the shell displacement in the radial and tangential directions, external pressure, internal pressure, and mass, respectively, and the subscript m indicates the m-th order of the oscillation. Moreover, A_{1m}, A_{2m}, A_{3m}, and A_{4m} are the coefficients of the m-th order expansion series that includes Young's modulus E and the membrane thickness h/a.

The internal pressure, Π_{int}, was derived from the relationship between the bulk modulus K and the cell volume. The mechanical properties of the intermediate filaments have non-linear stress strain curves in stretch tests because the pressure in the cell is thought to vary non-linearly with the cell volume. Therefore, the new K is

$$K = K_0(1 - w_m)^n \tag{3}$$

where K_0 and n are the initial bulk modulus and a non-linear parameter, respectively. In this study, w_m was much less than 1, so

$$K = K_0(1 - nw_m).$$

$$(4)$$

A value of $n = 0$ indicates that the structure has linear stress-strain properties and $K = K_0$.

If we apply the modified bulk modulus defined in Eq. (4) to the oscillating equations (1) and (2), we obtain non-linear oscillating equations for a cell with internal structures.

Figure 9 shows the predicted frequency response with different non-linear parameters n for two cells. The amplitude indicates the non-dimensional radial displacement of the oscillation. In this calculation, the membrane thickness parameter $h/a = 0.1$, initial bulk modulus $K_0 = 1000$ MPa, ratio of the two cell diameters $\varsigma = 1$, non-dimensional distance $d = 2.5$, and Young's modulus $E = 3.0$ MPa. As the non-linear parameter n increased, the amplitude of the second natural frequency varied to some extent. The stiffness of the cell increased rapidly with n, so that when $n = 100$, the amplitude at 8 MHz was greater. From this result, the differences caused by the internal structures may possibly affect the cell damage caused by shock waves.

5   Conclusions

Comparison of a hyperelastic solid cell model and a hyperelastic membrane with fluid-filled cytoplasm model showed that the former reproduced the cellular deformation during substrate stretching properly, while the latter predicted the opposite displacement of the cell membrane because of the assumption that the cytoplasm had no structural components. A cytoskeletal network model should be incorporated into the latter model to describe the transfer of the external forces between the apical and basal membranes, and between the nucleus and these membranes.

During freezing and thawing, the tissues showed remarkable morphological changes in their histological microstructures, depending on the thermal protocol. The patterns of the histological change were determined by the freezing pattern, which was dependent on the cooling rate. Moreover, the deformation and meandering of muscle fibers tended to be promoted by a slower warming process. These morphological changes in the histological microstructures affect the mechanical properties of the tissues.

After gathering experimental data on cell damage due to shock waves, the deformation process in each case was evaluated using a special mathematical model for two-cell deformation. A frequency response analysis demonstrated that the effects of the nonlinear parameter on the deformation process were large at certain frequencies. More information about the effects of the cell structure on the degree of damage is required, and may be obtained from both theoretical models and experiments.

Acknowledgements

This work was supported by Grant-in-Aid for Scientific Research on Priority Areas 15086213 from the Ministry of Education, Culture, Sports, Science and Technology of Japan.

References

1. Caille, N., Thoumine, O., Tardy, Y., Meister J.-J., 2002. Contribution of the nucleus to the mechanical cells, J. Biomech., 35, 177-187.
2. Yamada, H., Matsumura, J., 2004. Finite element analysis of the mechanical behavior of a vascular endothelial cell in culture under substrate stretch, Trans. Jpn. Soc. Mech. Eng., Ser. A, 70, 710-716 (in Japanese).
3. Yamada, H., Kageyama, D., Takahashi, Y., 2004. Modeling of an endothelial cell taking account of elasticity in cell membrane and fluid pressure in cytoplasm, Proc. 16th Bioeng. Conf., JSME, 17-18 (in Japanese).
4. Gage A.A., Baust, J., 1998. Mechanisms of tissue injury in cryosurgery, Cryobiology, 37, 171-186.
5. McGrath, J.J., 1993. Low temperature injury processes, Advances in Bioheat and Mass Transfer, ASME HTD-Vol. 268, 125-132.
6. Chaussy, Ch. et. al., 1982. Extracoporeal Shock Wave Lithotripsy, Karger.
7. Teshima, K., Ohshima, T., et al., 1995. Biomechanical effects of shock waves on Esherichia coli and λ-phage DNA, Shock Waves, 4(6), 19.
8. Ishiguro, H., Horimizu, T., Kataori, A., Kajigaya, H., 1999. Three-dimensional microstructure of biological tissues during freezing and thawing (Real-time observation by confocal laser scanning microscope), Trans. JSRAE, 16(3), 283-295.
9. Ishiguro, H., Horimizu, T., 2002. Three-dimensional behavior of ice crystals and cells during freezing and thawing of biological tissues, Proc. 12th Int. Heat Transfer Conf., 291-296.
10. Tamagawa, M., Akamatsu, T., 1999. Effects of shock waves on living tissue cells and its deformation process using a mathematical model, JSME Int. J. Ser. C, 42(3), 144.

SUBJECT INDEX